NATURE'S GREAT EVENTS

NATURE'S GREAT EVENTS

GENERAL EDITOR KAREN BASS

INTRODUCTION BY BRIAN LEITH

MITCHELL BEAZLEY

NATURE'S GREAT EVENTS

First published in Great Britain in 2009 by Mitchell Beazley,
an imprint of Octopus Publishing Group Limited,
2–4 Heron Quays, London E14 4JP
www.octopusbooks.co.uk

An Hachette Livre UK Company
www.hachettelivre.co.uk

By arrangement with the BBC

ISBN: 978 1 84533 4567

A CIP record for this book is available from the British Library

Commissioning Editor: Hannah Barnes-Murphy
General Editor: Karen Bass
Art Director: Tim Foster
Senior Art Editor: Juliette Norsworthy
Designer: Lizzie Ballantyne
Picture Research: Laura Barwick
Cartography and illustrations: Martin Darlison at Encompass Graphics
Project Editor: Georgina Atsiaris
Copyeditor: Caroline Taggart
Proofreader: Ruth Patrick
Indexer: Hilary Bird
Production Manager: Peter Hunt

Set in Quadraat and The Sans
Printed and bound By Toppan Printing Company, China

Contents

Introduction

It's startling to realize that the living world we see about us – all those plants and animals, all the forests, grasslands, jungles, and oceans, and even the seasons and the migrations that take place every year – the whole three-ring circus we know as life on Earth owes its unique character to a tiny cosmic accident. In the grand scheme of things – on the "universe" scale – this "accident" is laughably random. But for us mere mortals, scrabbling around for survival on the surface of our tiny world, it holds the key to all life.

That accident is the tilt of our planet. Pick up an orange and imagine it is Planet Earth. Hold the stem end at the top – north – with the navel dimple at the bottom – south – and imagine it spinning about an axis running from the stem through the centre to the navel. In a perfect universe that's how our world would be spinning: on an axis running vertically through the South and North Poles, with the sun 147 million km (92 million miles) away on the horizon. A day is a single spin of the globe around its own axis, a year the time it takes for the globe to travel around the sun.

Slowly spin the orange about its axis and imagine heat and light arriving from that distant sun. In this fruity universe, the orange's equator is always hot – because it's always facing the sun – with the rest growing gradually cooler as we move towards the poles. In this notional orange world, the sun's energy will always strike the equator first, no matter what time of year it is, and the northern and southern hemispheres will be warmed equally.

The trouble is – or, as we perhaps should say, our great good fortune lies in the fact that – we don't live on a perfect orange planet. Our world is lopsided. The axis running through it isn't vertical, but sits at an angle of about 23 degrees. If you tilt the orange in your hand a bit and slowly spin it on its axis – and again imagine heat and light striking it from an imaginary distant sun – you'll see straight away how that tilt changes everything: the sun's energy doesn't arrive at the equator first, but at the bulge of the orange just above or below it. It's exactly the same with our Earth: the tilt means that the planet is not uniformly heated: at certain times of the year (when the earth is on one side of the sun) the southern hemisphere "bulge" receives more of the sun's energy, and at other times (when the earth has moved around to the other side) the northern hemisphere gets more.

It's this lack of uniformity that creates the seasons. As we slowly circle the sun over the course of a year, the amount of heat and light reaching us varies from summer (maximum) to winter (minimum). Where I live – in Bristol, England – we sit roughly half way between the equator and the North Pole. In winter we get a measly nine hours of daylight a day, whereas in summer we get nearly 16 hours. That tiny tilt makes a huge practical difference to life in a northern latitude!

But the point is that these seasonal differences are responsible for just about everything interesting on our living planet.

The seasonal cycle – this sequential heating and cooling over the course of the year – drives vast swirls and eddies of oceanic and atmospheric currents around the world. It is responsible for annual patterns of freeze and thaw, rain and drought, growth and die-back. The tilt of the Earth means that the sun's heat reaches much further north in our summer than it ever would in a perfectly symmetrical world, thereby enabling life to thrive at surprisingly high latitudes; but it also means that in the depths of winter the North Pole is shrouded in darkness, while the South Pole basks in 24-hour-a-day light. And it's this tilt – and the resulting seasons – that are responsible for nature's great events.

Every year, the seasonal warming and cooling of the oceans, the atmosphere, and the great landmasses create huge cyclical crises – and opportunities – for hundreds and thousands of species. Tongues of warm-water currents lick deep into cold oceans; rain clouds dry up and move on, leaving drought; increased light levels and day length trigger a huge bloom of plankton – a feast for a billion herring. Some of these seasonal cycles wreak havoc, others bring plenty in their wake.

Imagine the Arctic. For several dark months during the northern winter, these cold latitudes become hostile to life. The temperature plummets below zero and ice locks both land and sea in a frozen, alien brutality. Then comes light and growing warmth through the spring and summer, the thawing land and sea emerge once more from their icy imprisonment. Opportunities abound: birds fly in to nest and breed; fish and whales swim northwards to feed; land animals such as caribou arrive to take advantage of the sudden bloom of grass and tundra, with predatory wolves and foxes not far behind. The seasonal thaw triggers the activity of thousands of species, from flies to foxes, drawn into a short-lived explosion of opportunity and danger.

And seasonal change can be just as dramatic on the equator itself.

We've all seen film of the famous Serengeti migration, where tens of thousands of zebras and ten times that number of wildebeest roam the grasslands of East Africa in search of fresh pastures. This, too, is a seasonal event – with a somewhat unpredictable pattern and schedule triggered by the alternate warming and cooling of the vast African landmass and the great reservoirs of thermal energy that are the neighbouring oceans.

As soon as rain starts to fall in the southern Serengeti, the animals leave Kenya's Masai Mara and head south into central Tanzania, driven by memories of richer grasses in ancestral birthing grounds. The most surprising thing about the migration is that this – possibly the world's most famous natural event, involving dozens of species and over one million animals – takes place in a circular route that could quite easily be contained within an area not much bigger than southern England.

Virtually all of the planet's most dramatic natural events – the great aggregations of life, the huge explosions of opportunity and danger, the coming together of predators and prey – are driven by the seasons. Just imagine what a boring world we would live in if we didn't have a tilt!

As wildlife television film-makers we saw a huge opportunity in telling the stories of these great events. If the success of Planet Earth has taught us anything, it's that there is a huge hunger out there for the dramatic: the Earth from the air, the grand overview of the world's great landscapes. Rarely can a single piece of camera technology – the famous "cineflex" aerial mount – have caused such a stir. Who could forget remarkable shots like the slow zoom-out to reveal one million snow geese flying south over the eastern seaboard of the USA; or the close-up of a wolf, from the air, finally running down and killing a caribou after a 19km (12 mile) chase? The great events we wanted to show offered huge opportunities for such epic visual moments. We booked the cineflex!

But it wasn't just the large-scale and the epic we wanted to reveal. Great dramas feature intriguing characters and intimate, personal stories too. If we saw an opportunity to go big and bold, we saw an equal opportunity to document the small and intimate – to tell stories of struggle, survival, and even death as experienced by the characters caught up in our natural dramas. From the very start of planning the series we thought: who are our main characters going to be? Through whose eyes will we see this action? Some of the most memorable wildlife films ever made have been stories of individual animals – a wild dog or a meerkat, an elephant or a bear – caught up in events much bigger than themselves, struggling to survive against all odds, or perhaps just lucky enough to escape with their lives despite overwhelming adversity. We even saw an opportunity to tell emotional and engaging stories about creatures that had been drawn into an event for their own very specific reasons, but that might end up facing each other in a life-or-death struggle to survive.

So, we had epic and we had intimate. But there was a final, key ingredient we were looking for in our film-making recipe: a strong narrative. There was a time – and not all that long ago – when wildlife films were often glorified biology lessons. Go to a place in Africa and film everything that moves. Choose a species and film all of its natural behaviours. Observe the breeding strategy of the lesser-spotted woodpecker... These weren't dramas, merely snapshots of biological reality, often without any structure or story-telling "arc". Today, the fashion has changed and audiences respond more favourably to strong, emotionally engaging stories. And if there's one thing our "great events" have in spades, it's the potential for strong narratives. From the beginning of production we were thinking about where our stories would start and end, making sure we established our key characters near the beginning, and that the events reached their climax near the end. Of course when we started filming we had no idea what would happen to our characters – or, indeed, what would happen to the events themselves – but we were on the look-out for personal dramas and we were lucky to have powerful "natural" narratives built into the films from the start.

And so we started our own epic adventure to film the world's great natural dramas. And what we discovered, is that our world is changing.

The experts told us that the Okavango River in Botswana floods in May as the rains from Angola flood into the arid delta. But the rains in Angola had not fallen on schedule, and local rains had been so abundant in the delta that a flood pattern that has been reliable for over 20 years was completely different. The gurus maintained that the annual "Sardine Run" off the south-east coast of South Africa would happen in late May or early June as the cold-water currents forced their way up the eastern coastline. But the gurus hadn't taken climate change into account. And of course we all now know that the summer thaw in the Arctic is happening earlier, and lasting much longer, than it used to – so that the region's polar bears face a very uncertain future without their winter ice.

In the end, therefore, we fulfilled another purpose quite apart from the epic, intimate, and narrative-driven stories we had set out to create: we became witnesses to, and chroniclers of, a fast-changing planet.

In a sense, the timing of our series could hardly have been worse. The natural great events we chose to film have probably never been more unpredictable, and harder to pin down and capture on film. But in another – much more important – sense, the timing could hardly have been better. If our series has helped to hasten a growing public awareness that our planet is undergoing rapid and serious climatic change, and that the seasonal cycles and events we've all grown up with are rapidly destabilizing and unravelling, then perhaps it may serve a very useful purpose, beyond mere entertainment.

Brian Leith, Executive Producer

Foreword

The television series on which this book is based follows six of the most dramatic wildlife spectacles on Earth. Our mission as a team was to film these natural events. Each is triggered by seasonal change on a large scale, whether it's the flood of the Okavango Delta in Africa, the melt of the Arctic ice, or the massive bloom of plankton in the northern Pacific. These seasonal extremes fundamentally transform the environment, directing the movements of huge numbers of animals.

The locations are stunning wild areas and one of our tasks was to show how they change through the year. Filming weather, often wild and dramatic, revealed the processes of change, while a combination of beautiful aerial filming and Earth-from-space graphics showed the scale of transformation. But key to the emotional appeal of the films were the dramatic lives of the animals caught up in these events. Inevitably there are winners and losers, and often surprises. The sad sight of a family of lions starving, while the herd of wildebeest are many miles away on their epic migration, turned the stereotype of Serengeti wildlife on its head. Many of these animals struggle against the odds. Salmon heading home up the rivers of British Columbia, for example, have to dodge hungry grizzlies and wolves and fight their way up huge waterfalls just to make it to their spawning grounds.

To achieve the ambition of this kind of series needed an experienced team from the BBC's Natural History Unit in Bristol, working with wildlife filming experts on location all around the world. Each member of the team had his or her own expertise, from underwater filming to polar survival, backed up by experts in production management and logistics. We also had a world-class team of specialist wildlife photographers, many with detailed knowledge of their locations. We also relied heavily on the support and knowledge of local people, scientists, and wildlife experts who live and work permanently in these remote locations.

From the polar north to the tropics, our teams ventured into some of the planet's most breathtaking wildernesses. As we embarked on the series, it soon became apparent just what a challenge it would become to capture fully these great natural events. With global climate change, seasonal timings are becoming harder to predict. For us filmmakers a fast-changing, unpredictable planet is a major headache. For a hungry polar bear waiting for the ice to refreeze so he can hunt again it's a matter of life and death.

Even with the best information, based on scientific data and local knowledge, being in the right place at the right time proved to be our biggest challenge. The herring spawnings off Alaska in 2007 and 2008 were the latest recorded for 25 years. During that same period, the flood of the Okavango Delta was the most extensive it had been in 8 years. A good thing, you might think, except that the water dispersed some of our animal characters over such a huge area that they were impossible to find for months. And in 2007 the sardine run simply did not happen at all. There was nothing we could do but wait. It was only the following year, during the final few months of production, when editing had already begun, that at long last a current of cold water pushed north along the African coast, bringing with it one of the most action-packed wildlife spectacles on earth.

We had 25 months from research to delivery of the completed programmes. For a series like this, based on seasonal events, that isn't long. When combined with the unreliability of the events themselves, it certainly took our schedules to the wire. However, after 1300 days filming spread over 18 months, and weeks spent planning and plotting, we finally completed our mission.

The Earth is changing fast; making this series was a sharp reminder that we cannot take these natural events for granted. Working on the project and witnessing these changes first hand sometimes felt like taking the pulse of the planet itself. In celebrating the wonder, complexity, and spectacle of nature's great events, all of us on the team have also come to appreciate their fragility, and the vulnerability of the wildlife that is dependent on them.

Karen Bass, Series Producer

Event locations

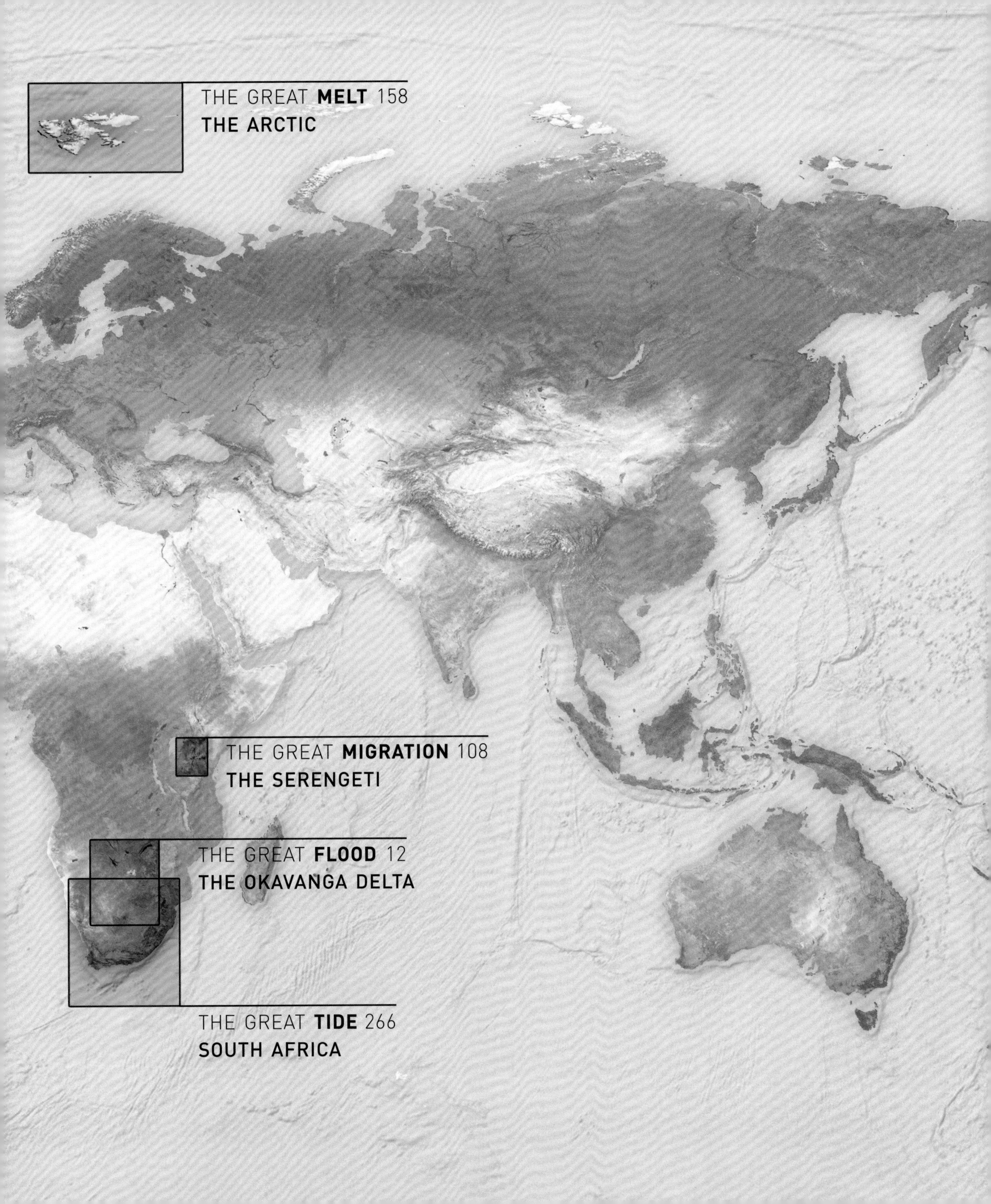
THE GREAT **MELT** 158
THE ARCTIC
THE GREAT **MIGRATION** 108
THE SERENGETI
THE GREAT **FLOOD** 12
THE OKAVANGA DELTA
THE GREAT **TIDE** 266
SOUTH AFRICA

The Great Flood
The Okavango Delta

If you gaze down from space, high above Planet Earth over the Tropic of Capricorn, the vast Kalahari Desert unfurls below you across an immense swathe of southern Africa. An ancient wind-blown coat of sand dunes – cadmium orange and ivory white and 65 million years old – rolls like an ocean swell across the mountainless land, blanketed in a velvet mantle of hardy grasses and stunted trees. At its heart, amid the desert sands of north-eastern Botswana, a luxuriant emerald oasis spreads a giant hand, its watery fingers reaching 200km (125 miles) into the parched Kalahari. This is the Okavango, "the land of many rivers", one of the largest and certainly the most pristine inland freshwater delta on earth.

Every year in June, as the dry African winter season grips the desiccated land, and temperatures tilt towards zero, a miracle occurs. The waters of the Cuito and Cubango, two great Angolan rivers that feed this delta, never reach the sea. Instead they merge to form the Okavango River which, as it enters Botswana, disperses in slow, serpentine motion across 16,000sq km (6000 square miles) of scorched Kalahari sand, fragmenting into hundreds of small waterways to create a wildlife paradise unequalled in all of Africa. Billions of litres of crystal-clear amber liquid surge down narrow channels, fill glittering lagoons, and sheet thinly onto parched floodplains of yellow grass – a rejuvenating pulse of nourishment to which all of nature quickens. This, one of Earth's truly great natural events, is the Okavango flood.

The Okavango Delta
Botswana

Botswana is a landlocked country about the size of France, nestled amid the Kalahari Desert in southern Africa. In the north-east of the country lies the Okavango, the "jewel of the Kalahari" – a vast, pristine, seasonal swampland that is one of the most important wetland habitats on Earth.

Although much of Botswana's wealth is based on diamonds and the cattle industry, tourism is an increasingly important resource. Many visitors enjoy the Okavango every year, not only for its spectacular scenery and dazzling wildlife, but also because the desert and delta offer the genuine opportunity to spend time in one of Africa's last remaining wild and unspoiled wilderness areas.

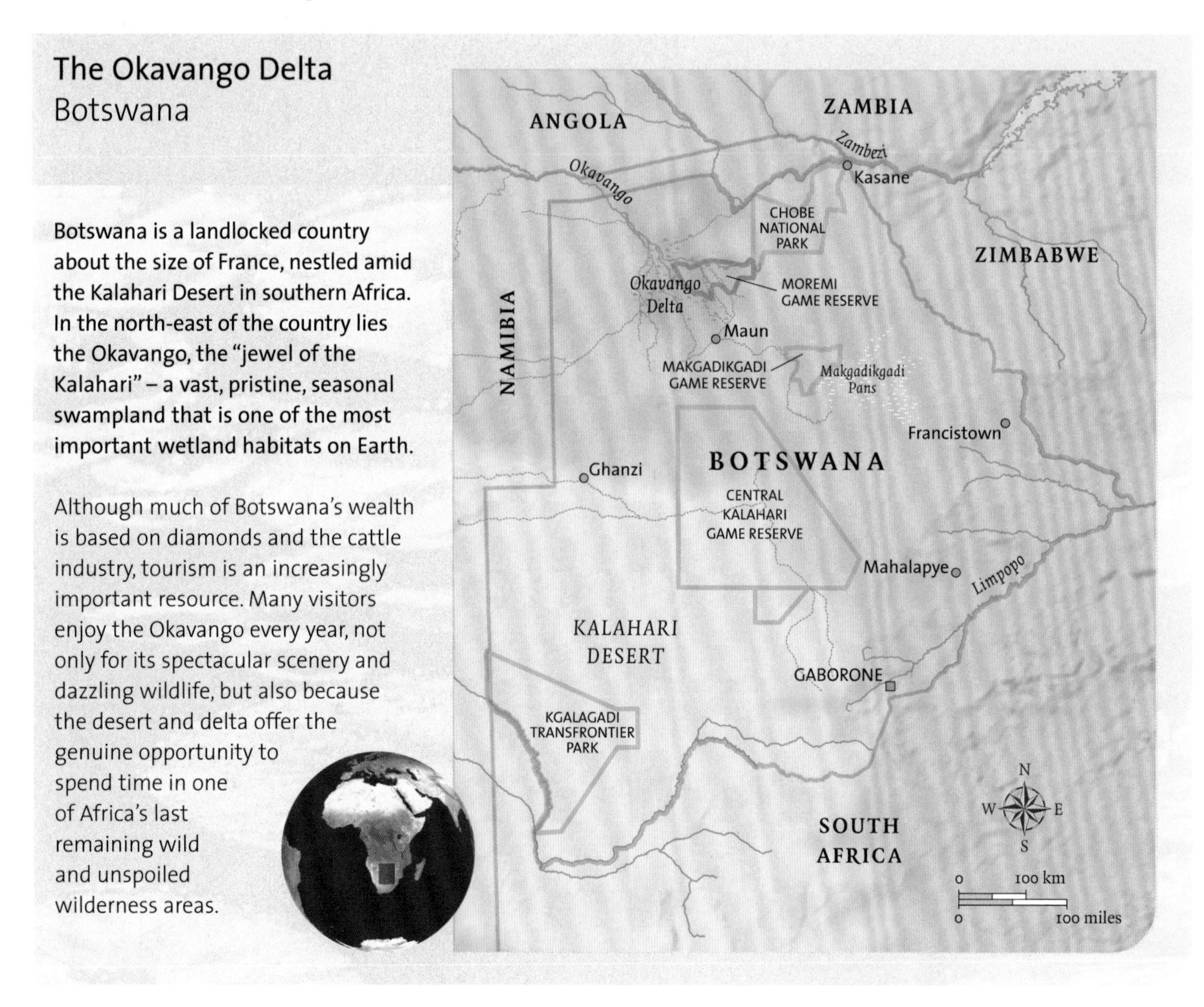

Xigera Lagoon (right): a spectacular "delta within the delta" that almost perfectly mirrors the shape and pattern of the Okavango.

Little bee-eaters (above): a group huddles together for safety at night. The Okavango is a bird-watcher's paradise, home to over 440 species of birds.

What is "the flood"?

The great paradox of the Okavango is this: at the height of winter, when the 45°C (113°F) heat and torrential summer rain are a distant memory and the land has dried to extremes, the flood arrives. The secret lies in Angola, which receives its limited annual rain (mostly in the form of tropical thunderstorms) in the summer months, from November to March. This takes six months to flow 1000km (625 miles) to reach the Okavango, and by this time, in June or July, the delta is at its driest. As the last delta grasses wither and the remaining pools parch into rock-hard mud, the flood truly does seem like a miracle.

The flood itself is a complex event, primarily because it is so unpredictable. Its height and extent vary every year, in better years covering more than double the area (about 12,000sq km/4500 square miles) of lesser years (about 5000sq km/2000 square miles). Its size and dispersion are determined by three factors: the summer rainfall in the Angolan catchment area; local rains in the delta itself; and the remaining groundwater from last year's flood, and thus the local water-table level as the new flood arrives. Generally, in the upper delta, water levels begin to increase in March, and swell until the flood reaches its greatest extent in August, when the floodwaters finally arrive at the most southerly reaches of the delta. By this time, levels in the upper delta have already begun to fall. Layered upon this basic pattern are innumerable variables such as shifting channels, papyrus blockages, and important hippo and termite activity – variables that add such levels of subtlety and complexity as to challenge comprehension.

The result of all of this complexity is an astonishing diversity of both island and wetland habitats, which support the spectacular array of insects, amphibians, fish, birds, and mammals for which the Okavango is renowned. There are 122 species of mammal, over 440 birds (including 38 species of eagles, hawks, buzzards, and kites), around 100 different reptiles and amphibians, over 70 kinds of fish and about 5000 varieties of insects. Yet despite these species counts and the enormous variety of habitat, numbers and concentrations of these creatures, with a few exceptions, are perhaps less than one might imagine. This is, after all, a special corner of the Kalahari, where nutrients are generally low, the carrying capacity of the grasses limited, rainfall scarce, and plant communities less productive than their counterparts elsewhere in Africa.

Despite this, however, the Okavango is home to some of the last great herds of buffalo in Africa (herds numbering 3000 are not unusual) with an estimated population in the delta of more than 20,000. There are 35,000 elephants resident in the Okavango region, a substantial proportion of

Architects of the delta (left): hippos play an important role in keeping open the many channels and waterways as they wander back and forth in search of grazing grounds.

Botswana's estimated population of 120,000. There are several thousand hippo, a significant world population of lions (1500) and probably well over a thousand African wild dogs, a sizeable percentage of the continent's remaining population. A number of swamp specialists live here too – red lechwe, large aquatic antelopes characteristic of the Okavango, number an estimated 40,000, and sitatunga, secretive antelopes that live their life almost entirely confined to the papyrus and reed beds, are thought to number 5000, although this is probably an underestimate. About 15 per cent of the world population of the rare and majestic wattled crane breed here, and among the extravagant array of bird species, the slaty egret uses the Okavango as one of its few remaining breeding sites. A kaleidoscope of raptors, rollers, storks, hornbills, waterfowl, kingfishers, bee-eaters, and many others make the Okavango one of the greatest bird shows anywhere on our planet.

A fragile paradise

Although the Okavango is a highly dynamic complex of rich and fecund ecosystems, it is also extremely fragile. Owing its very existence to the vagaries of the annual flood, the delta is at the mercy of changing catchment area rainfall, local rains, massive evaporation, desiccation by wind, assault by fire. Yet every year, to a greater or lesser degree, the delta is rejuvenated by this one great event.

A staggering 10 cubic kilometres (10 trillion litres or 2.2 trillion gallons!) of water flow into the system each season. Perhaps most remarkably, 98 per cent of the water that pours into the delta never flows out – it is all lost to transpiration or evaporation. More noteworthy still, thousands of tonnes of deposits, and particularly salts, are assimilated by the delta every year, and yet the entire system manages to survive the potentially lethal salt poisoning. How it does so is an obscure natural phenomenon that gives the Okavango all the outward signs of being a highly intricate, complex, living organism.

Nestled in the north-west wilderness corner of Botswana known as Ngamiland, the Okavango is a place of constant change, and has been so for aeons. The delta today is a sparkling memory of a once vast ancient lake which once covered 155,000sq km (60,000 square miles), but which has progressively contracted over the last 2 million to 25,000 years. More than that, however, it is the product of a natural geological phenomenon – a series of unlikely fault lines that are the distant, most southerly shudderings of the Great Rift Valley. At its northern extreme, the incoming Okavango River is constrained by an invisible subterranean chicane of rock, its walls formed of two parallel fault lines, creating an area known locally as the Panhandle.

African paradise (right): fields of lilies adorn the many quiet backwaters and lagoons of the Okavango.

Tiger fish (above): these speedy and voracious predators haunt the fast-flowing channels, preying on smaller fish, frogs, and even small birds.

Under and above the water (above): the waters of the Okavango carry little sediment and are crystal clear for most of the year. The water has a characteristic pale golden colour, the result of dissolved iron and fine organic material.

This narrow, 80km (50 mile) long corridor is dominated by papyrus and bisected by the lazy meanderings and oxbows of a single big river.

Downstream, the fault lines release the river from captivity, and as it shrugs off its rocky bonds, the water begins to spread, breaking into a fan of channels that percolate out across the sand. This great basin is almost flat, and filled 300m (1000ft) deep with the sand deposits of 65 million years. Remarkably, the gradient of the delta is a mere 1:3,300 – a decline of only 60m (200ft) over its 250km (150 mile) length. It is this incredibly shallow profile that encourages the incoming river to spread so widely. At its broadest, most southerly span, the delta is 160km (100 miles) wide, and here reveals its final geological secret – two arrow-straight fault lines cut perpendicular to the water's gentle flow, forming a natural subterranean dam that gives the Okavango its dramatic and distinctive shape.

From the air, the Okavango is an exquisite and intricate tapestry of cerulean and turquoise channels, viridian and ochre floodplains, and cobalt lagoons, sprinkled with over 150,000 emerald islands that vary in size from a few metres to many kilometres across. This mosaic is constantly shifting. Thousands of tonnes of sand are carried in by the flood every year. As the water slows across the great alluvial fan, those sands are deposited in the main channels and tributaries, giving rise to a complex system of "channel switching" that alters the flows and the face of the delta every year.

"Okavango champagne" – the purest water on Earth

One of the most characteristic features of the Okavango is the astounding purity and clarity of the water. Upstream there are few sources of pollution, but more importantly, the soils of the catchment area are not easily eroded, and consist mostly of sand rather than fine mud, keeping the water clear. Yet this water clarity also means that the inflow to the Okavango is very low in nutrients, which raises another conundrum – if nutrient levels are so low, why does the Okavango enjoy such abundance and diversity of plant and animal life? The answer is almost biblical: earth, wind, and fire. Firstly, 500,000 tonnes of dissolved mineral and plant material have washed into the delta annually for thousands of years, accumulating over time. Then, an estimated 250,000 tonnes of nutrient-rich atmospheric deposits rain in or are blown in during the winter. In addition, every year, large stretches of the swamp actually burn and this releases vital nutrients into the system. These three sources are richly supplemented by the local decay of aquatic and terrestrial plant material, and by hippo and elephant dung.

As flood turns desert into swampland every year, most of the Okavango's 330,000 large mammals have to deal with life in shallow water. For water-loving animals – hippo, lechwe, sitatunga, crocodile, otters, and even the great herds of elephant – this is a paradise. But for dry-land species, the flood presents a unique set of challenges. The availability of island habitat on which these animals depend, whether to graze, browse, or hunt, becomes severely reduced almost overnight. Animals are forced to wade through water as they move between feeding areas. Home ranges become compressed, forcing territorial species like lions to adapt their behaviour. And vast expanses of dry, sandy, grassland are drowned, forcing millions of small mammals and insects to take refuge on the islands.

Undoubtedly the unique spectacle of the Okavango is seeing the familiar savannah animals of Africa – elephant, lion, cheetah, hyaena, zebra, wildebeest, giraffe, and a host of smaller mammals – coping with such unusual conditions. The true magic of the delta lies in this great variety and proliferation of animals and birds, set against the magnificent backdrop of waterways and islands and flooded grasslands. There is nowhere quite like it on Earth.

Below the surface (right): a dazzling array of aquatic plants fills the clear channels and lagoons, providing a diversity of ecosystems that are home to an equally impressive number of aquatic animals.

Anatomy of a swamp
Levee system and channel switching

The Okavango creates its own unique system of levees as it grows and shifts, leading to the surprising and counterintuitive result that the riverbeds are often higher than the surrounding floodplains. How does this work?

One of the key factors that determines the location of islands and the spread of water in the Okavango is the channel-switching mechanism. It is because of this that the Okavango exists as a spread-out inland delta instead of a single meandering river.

Although the waters of the Okavango River are not very rich in dissolved nutrients and suspended particles, they do carry suspended salt and sand particles. As the water enters the relatively flat delta, its flow spreads out and slows down. The first plants to make use of the water coming into the delta are the papyrus, found along the edges of the deeper channels. The papyrus beds are extremely fast-growing and rapidly utilize the dissolved nutrients, converting them into peat beds that line the channels.

As the water flow slows down, dissolved sands are deposited on the riverbed, causing it to rise. But the rising beds of peat lining the sides of the channel keep the water in the centre relatively deep. These processes result in the whole river becoming slightly raised in relation to the surrounding floodplains. The water flowing out through the papyrus beds to the lower-lying land triggers an increased flow of nutrients into the papyrus beds. This fosters growth and the beds become more and more dense until, sometimes, they completely block the flow of the channel. The river then breaks out sideways and forms a new channel wherever it can, often along a path created by the movement of hippos.

Papyrus (above): this common plant grows in profusion in the upper reaches of the delta, and is an important part of the Okavango's natural filtering system.

The system of levees helps restrict the build-up of salt concentrations in any particular area, as the annual floodwaters continually change their route before ending their flow in the Kalahari sands. Eventually, when a channel has been diverted, and the papyrus died due to lack of sufficient water, the levee of raised nutrient-rich soil is all that is left. This then results in the formation of long, narrow, winding islands that ultimately become covered in dense riverine forest.

A mother hippo and her calf (below) travel down one of the sandy channels that form a network of waterways between deeper lagoons.

Satellite images of the Okavango flood

Before the coming of the flood (above): the Okavango viewed from space in April. The delta's fan shape is clearly seen, with the Okavango River entering it in the north-east (top).

Four months later (above): the same perspective from space shows the extent of the flood in August, when as much as 16,000sq km (over 6000 square miles) of the Okavango may be covered in water.

Permanent swamp

Between the Panhandle area and the seasonal floodplains lies an area of permanent swamp, covering about 4000sq km (1500sq miles). Characterized by deep channels, lily-covered lagoons, papyrus, reeds, bulrushes, and a host of aquatic plants, this permanently inundated area of the Okavango is home to hippos, sitatunga, crocodiles, otters, and water mongooses, as well as many bird species, frogs, and aquatic insects. There are thousands of small islands, many covered with wild date palms, and the waterways are lined with distinctive water figs. As the annual flood arrives, this permanent swampland acts like a giant sponge, absorbing the incoming water for a few weeks, until finally it breaks out into the main channels and seasonal floodplains.

Painted reed frog (right): one of many exciting species of amphibians that live in the important Ramsar wetland. Frogs are particularly sensitive to environmental degradation, and this unspoilt habitat is essential to their survival.

Fault lines

—— Fault line

Geology at work (above): the distinctive shape of the delta is created by a series of fault lines, geological shifts that are at the southern extreme of the Great Rift Valley. Two parallel fault lines confine the river to the Panhandle in the north-west. Passing these, the floodwaters spread into a vast dendritic fan. At the south-eastern fringe, another series of faults forms a natural subterranean dam.

Safari with a snorkel
Mike Holding

This particular story of the great flood in the Okavango begins, perhaps surprisingly, far from the delta, 100km (60 miles) into the desert. Of all the animals that take advantage of the delta's abundance during the flood, African elephants are one of the more charismatic.

Elephants travel immense distances in this part of the continent, and the populations that spend most of their time in the Okavango are no exception. When the season's flood is over, and the lush green floodplain grasses on which they feed have yielded to the battering assault of October's summer heat, thousands of elephants begin a trek into the dry mopane woodlands to the north-east, following "elephant highways" that are visible from space and have been etched into the landscape by the soft tread of countless generations. If the elephants' timing is right, rains will begin here in late October, and hundreds of small clay depressions known as pans will fill with fresh rainwater. Within days of the first rain, the gaunt, stunted mopane trees burst into life and a miraculous emergence of new leaves flushes the woodlands with a fusion of pale green, red, and ochre.

Mike Holding (above): a qualified zoologist and passionate wildlife filmmaker who has spent 20 years filming in the Okavango.

Desperate for water (left): during the summer rain season, elephants drink from isolated clay-lined waterholes known as "pans" in the mopane woodland to the north-east of the delta, often trekking many kilometres from one pan to the next.

Calmer waters (above): a young elephant follows its herd from a drinking pool. In contrast to their playfulness at the mud wallows, elephants tend to behave more sedately in these waterholes, to avoid muddying their drinking water.

For a few short weeks this is elephant paradise. The elephants sense every rainstorm, and follow highly localized downpours that soak only a few square kilometres for a few minutes. They jostle at the freshest waterholes, feed on the renewed luxuriance of the mopane trees, and then move on.

But this year, the rains are late. We have driven off-road in deep, hot sand for about 100km (60 miles), following fresh spoor along a well-worn elephant path, crashing through the almost impenetrable mopane forest in our camera vehicle, hoping to find a small waterhole that has enough remaining moisture for the elephants to slake their thirst. Finally, after two days of driving, we find a small pan surrounded by yesterday's elephant tracks, and settle down to wait. Mud oozes like liquid chocolate in the desiccated waterhole, hardly bigger than a few strides wide and covered by a thin film of grey-green water. Around us stand skeletal mopane trees, and infinities of sun-bleached sand, the white and sienna tones of the dry Kalahari.

The land has been beaten into submission by the stagnant heat. Eddies and swirls of mirage mock our attempts to scan for life through the binoculars; the angular trees of the horizon are warped into molten liquid forms by furnace air that engulfs the bush. It's 50°C (122°F) in the shade and the country smells of burning. Even the flies are indolent. In Maun, the dusty frontier town we call home, October is known as suicide month.

We shift restlessly, crouched low and seeking scraps of shifting shade beneath the leafless, raw-boned branches of an old mopane tree. We wait, drawing light and shallow breaths so as not to burn our throats. The metal of the vehicle is too hot to touch. And then, after hours of waiting, we see them, oily distorted shapes in the distant mirage – elephants – rolling like old galleons on an undulating sea, moving with great slow-motion strides down an ancient path to the waterhole.

Thirsty elephants usually rush headlong into water and splash with abandon, but this group does something I have never seen before – they step in very slowly, making as little disturbance as possible, mothers reprimanding unruly youngsters lest they churn up the water. And then remarkably, one after another, the elephants begin to carefully sweep their trunk tips across the surface, delicately siphoning the few centimetres of clear liquid from the mud below. I roll and roll on the camera, capturing the details of this elaborate and intelligent behaviour.

We stay at this receding waterhole for a few days, eventually filming dozens of breeding herds passing through, many of them repeating the extraordinary siphoning we had witnessed on the first morning. To our great delight, one young female arrives with a newborn bull calf only a few hours old, and I film his confusion and excitement as he experiences water

Elephant
Loxodonta africana

Botswana has an estimated elephant population of between 120,000 and 150,000, the largest contiguous population in Africa. Of these about 30,000 live for at least part of the year in the Okavango.

During the summer, when the delta is dry, elephants move north and east into the mopane woodlands and the surrounding Kalahari, feeding on mopane trees and finding water in the many perennial clay pans filled by summer rain. As winter approaches, and the pans dry to mud, the elephants begin their journey back to the delta, anticipating the arrival of the flood. Here their feeding patterns change – initially they forage on the many species of palatable delta trees and shrubs and then, as the floodwaters rejuvenate the grasslands, they turn to grazing on the floodplains, the bulk of their food coming from the sweet grasses nourished by the incoming water.

KEY FACTS

Weight males up to 6000kg (13,200lb), females up to 3250kg (7150lb), making the elephant the largest land mammal on Earth.

Height normally 2.7–3.2m (9ft–10ft 6in), some males appreciably larger.

Lifespan 60 years

Tusks both males and females have tusks, which continue to grow throughout an elephant's life and have been known to reach 3.5m (11ft 6in) and weigh over 100kg (220lb).

A favourite wallow (left): a breeding herd of elephants swim and play in the muddy waters of a drying pan. The elephants use pools such as this to cool themselves from the searing heat, and the dried mud helps to repel biting insects.

for the first time, constantly slipping and falling in the mud and plunging his little face into the unfamiliar liquid.

Suddenly one night, a great tropical thunderstorm descends upon us with all its fury; lightning splinters the sky, boiling dark clouds roll overhead, wind whips furiously at our tents, and heavy, corpulent raindrops beat upon us as we scurry to cover all our gear. Although the rain provides welcome relief from the oppressive heat, now everything is soaked, and we spend the dark night sorting and drying sopping-wet equipment and sleeping bags. By morning, with beads of rain still glittering on the grey mopane branches, our pool and many others for miles around are full. Within hours the trees will begin to leaf. The elephants have dispersed – thousands of them vanished overnight, following other distant thunderstorms. It is time for us to leave. As we retrace our tracks, we come across dozens of replenished pans and, to our amazement, hippos wallowing contentedly in one of them. Incredibly, they have anticipated the rain, and walked 50km (30 miles) from the delta in a few days to reach this pool.

Thundering skies – the bringer of life

For the next month, the sky heckles the earth. Every afternoon, we marvel at theatrical displays of towering cumulus cloud, piling in monumental heaps on the skyline, reflecting hues of purple, crimson, and carmine, as sunset approaches. Yet mostly these demonstrations are little more than hot air – the rain, for which the earth now thirsts, evaporates before reaching the ground. Occasionally an extravagance of cloud builds momentum, and then in the distance, we watch fine grey veils of showers falling in small isolated patches.

Finally, late in November, the sky concedes defeat and the swollen clouds relinquish their precious cargo to the land. Day after sweltering day it rains, and we retreat from the bush to our base in Maun, using the coming months to overhaul equipment, repair our severely battered camera vehicles, and plan logistics for the coming filming season.

The land revives with spectacular rapidity. Suddenly the bush is luxuriant and flourishing, the grasses shoulder high, and the air filled with the cacophony of a million insects. Wildflowers jostle for sunlight between

Summer thunderstorm (above): spectacular storms like this revive the thirsty desert, and elephants follow the storms by the thousand, seeking fresh water.

Wattled cranes feed in the shallows (above): the Okavango is the largest remaining breeding site for these elegant endangered birds.

a profusion of exuberant shrubs, and the familiar calls of favourite summer visitors – the yellow-billed kite and extravagant woodland kingfisher – accompany us everywhere. The deafening shrill of cicadas pervades the mopane woodlands. It is a time of rebirth; tiny newborn impala skip on quivering, matchstick-thin legs, huddling in nurseries for safety, their soft doe-eyes gleaming with amazement at their abundant new world. In the Panhandle, miniature crocodiles hatch from egg clutches on silver sandbanks, their giant prehistoric mothers shepherding them with remarkable maternal diligence. Quelea swarm in millions, nesting in massive noisy colonies and smudging the skyline in huge pulsating flocks that tumble like smoke. Trees all over the delta are in fruit, and birds, monkeys, and a host of other creatures savour this time of plenty.

The changing season

By March, tall grasses lean low with the wind and finally succumb to gravity, tilting to rest on the drying earth as the last summer heat wicks the remaining moisture from the land, the humidity of the last few months evaporating into the haze. Crisp dawn mornings now, we shrug on fleeces against the chill. The last rain pools are drying, stranding fish that have missed their cue and can now only hope that their evaporating puddles outlast the drying wind, until the flood returns. If it does. Pelican fleets trawl the shrinking pools in huddled groups, working in unison to corral small fish; saddle-billed storks stride with exaggerated steps, their bills sweeping for frogs; and egrets stand motionless at the pool fringes, waiting to strike. The bush is quiet, and many animals are still dispersed in their summer ranges. Day by day, the pervading buzz of the flush of summer insects fades.

Later in the month, we head out to Gomoti, our small but comfortable tented bush camp that serves as a year-round base. Nestled amid towering ebony trees, and built around a colossal and extraordinary sycamore fig from which the site takes its name, this remote and wild film camp on the Gomoti River lies within one of the Okavango's most magnificent and varied wildlife safari concessions, Chitabe. Their two luxury lodges are only an hour's drive away, yet we hardly ever meet their safari vehicles – at any time there are perhaps fewer than 30 people in a wilderness area of thousands of square kilometres.

Although Chitabe's wildlife activity is excellent all year round, it is particularly abundant from April until October, and the variety of dry island and swamp habitats, pristine wilderness, and exciting wildlife activity here make it a privileged place to do much of our filming.

I take off at sunrise from a short rough bush airstrip in our small Cessna 182, a light aircraft that is perhaps our most valuable tool. The incredibly complex terrain and network of waterways of the delta make ground travel exceedingly slow and often risky, and the plane provides the perfect way of doing reconnaissance. I spend many hours a year in the air, tracking the progress of the flood, finding animal concentrations, mapping water-crossings and routes on the GPS, and filming aerial scenes. This saves enormous amounts of time and drowned vehicles – I can accomplish in a few minutes what would take us weeks on the ground – locating animals and routes and river crossings, which we then follow up in the camera cars. On this occasion, I recce the Gomoti's floodplains, and as

Aquatic antelope (left above): male red lechwe bound effortlessly through the shallows. Lechwe occur throughout the Okavango, and are well adapted to living in a flooded environment.

Not everyone likes the water (left below): other African mammals are less well adapted than the lechwe to a watery lifestyle. Baboons are forced to cross water as they move between islands to feed.

the gold and emerald grasslands and clear channels unfold beneath the wing, it's evident that the surface rainwater of the summer season is rapidly evaporating. The main river channel is drying, and only a few pools remain. I bank steeply over a spot we call Hippo Island; the receding water has forced the local hippos to gather here, and I count more than 40 of them grunting and jostling for space in the diminishing pool. Later in the day we head out there in the camera vehicle, an off-road journey through rivers, grasslands, and floodplains that takes more than two hours, even though Hippo Island is only 4km (2½ miles) from our camp.

For the next six months we will return to this pool often, recording the fate of the hippo pod as the encroaching dry season reduces their pool and their grazing. Tempers flare as conditions worsen: we film a violent battle between the resident male and his challenger, as both enormous animals rear up on their hind legs and lock jaws in a titanic show of strength. There is a sickening crack as the dominant male breaks one of his massive teeth, but the challenger is repulsed and wades off to a nearby island to nurse wounds that could, if they become infected, bring about his end.

As the winter season begins to unfold, we head out daily before dawn with the camera, exploring the empty river channel, drying grasslands, and surrounding woodlands, filming the small stories of the desiccating onset of winter. But these forays and excursions serve equally as a time of winding up to the main season – it's a good time for testing equipment, modifying and improving the vehicles, and for me, re-establishing that mysterious rhythm of working with the camera, a finessing of skills that always takes a few weeks, before I feel confident that the workflow is seamless and organic and intuitive. We all know that now the waiting begins.

Hippo
Hippopotamus amphibious

Papyrus and reeds are common in the permanent swamp, and large accumulations of dislodged or decaying plant material may actually block the Okavango's channels. Hippos, the bulldozer gang, continually travel up and down these channels, and wander out to graze on islands or grassy plains at night. Their repeated movement both maintains old waterways and creates new channels. The floodwater, always following the path of least resistance, flows down the dusty pathways they create, and thus the bulldozer gang has an important influence on which areas will receive new floodwater from year to year.

The bulldozer gang is active mainly in the rivers, lagoons, channels, and floodplains of the Okavango. But during the rainy season, when thunderstorms drench the parched land and new grasses spring to life, hippos often wander tens of kilometres from permanent water, feeding under the stars, and then taking shelter from the sweltering heat of the day in pans – small clay-lined depressions away from the swampland that fill with rainwater. Their movement back and forth, taking a few kilograms of mud with them each day and spreading it around as they walk, enlarges the pans year by year.

KEY FACTS
Weight males up to 3000kg (6600lb), females to 2000kg (4400lb).
Height around 1.5m (5ft) at the shoulder.
Lifespan 35 years

Titanic clashes (above): hippo bulls battle for territory and the right to breed. Such disputes generally escalate before the flood arrives, as water levels drop and the few remaining pools – and their resident pods of females – are contested by the bulls.

Hippo families (right): from the air, the delta reveals many thousands of hippos, grouped into small family pods. The hippos spend their days resting and sunning themselves in quiet lagoons and backwaters, and emerge at night to feed on the grasslands.

Distant rain – the life pulse of the flood

In the southern highlands of Angola, 1000km (625 miles) to the north-west, the rains have been late, but torrential. I have been on the internet, linked to cyberspace by sat-phone, watching the colourful shifting patterns of high-tech satellite images which divulge the weather's distant secrets every few hours. A lot of rain in Angola. I imagine that deluge gathering, roiling into eroded gullies and rushing down highland valleys to the Cuito River, and the Cubango. For the delta, these are the rivers of life. Over the next weeks they will swell into full pregnancy, carrying millions of tonnes of precious liquid, joining in a swirl of blue and brown eddies to form the Okavango River, 160km (100 miles) west of Botswana. This is our life blood, but it will be months before that first heartbeat is felt in the main delta.

As the floodwaters rise in the Panhandle, the area becomes alive with activity. We film hundreds of gorgeous carmine bee-eaters nesting in the few high banks of the river and watch skimmers valiantly fighting off the marauding attacks of fish eagles, which have a voracious eye for small skimmer chicks. The nestlings are equally at risk from crocodiles, large, water monitor lizards, and even keen-eyed goliath herons, and the skimmer parents repulse these attacks while desperately trying to raise their chicks before the floodwater covers their sandbank nests. The broad mats of papyrus float upwards on the rising tide, and a vast cavernous world becomes available for fish and small aquatic creatures under its protective mantle.

From the Panhandle, the floodwater begins to infiltrate the 4000sq km (1500 square miles) of permanent swamp. This is a watery place of tall reeds, sedges, bulrushes, papyrus, and small islands, inhabited by sitatunga, hippo, the occasional bull elephant, and a host of resident waterbirds. For a few weeks, the flood becomes practically invisible as it gently inundates this area, which appears to act like an enormous sponge, absorbing the first new pulse of water. And then, almost overnight, the flood breaks out of the permanent swamp and begins coursing into the

Carmine bee-eaters (above): thousands of these eye-catching birds nest in the steep banks of the river in the Panhandle area of the Okavango as the floodwaters begin to arrive.

Up the lazy river (right): the Okavango meanders through a vast swathe of papyrus, flowing in leisurely arcs through the Panhandle before spreading into the main delta.

Termites *Macrotermis michaelseni*

The Okavango is sprinkled with thousands of islands – some only 1m (3ft) wide, others many kilometres long. Many of them owe their existence to the humble mound-building termite.

When the floods have receded, and the lush grasses of the rainy season are abundant, a colony sends out thousands of alates – the winged reproductive form of the termite. These fly out in huge swarms, just after the summer rains, in search of new pastures. Once a male and female find each other, they drop their wings, dig a hole, and prepare to start their own colony. If, despite a large variety of predators, heavy rains, high flood levels or lack of sufficient food, the new colony manages to survive and start building its mound above surrounding lands, there is a good chance that it will continue to grow for up to 80 years.

Termites collect dead wood or grass and partially digest it. The resulting faecal pellets are then deposited in a central chamber in the mound, where a symbiotic fungus converts them into a food source for the termites. They keep the conditions in the mound perfect for the fungus by opening and closing holes like a primitive air-conditioning system. The mounds are also rich in nutrients, as they are built up of clay, sand, and mineral salts, and contain large quantities of organic matter, as well as the fungal gardens farmed by the termites. As a result they are quickly colonized by pioneer plant species, and small islands begin to form. Thus not only are termites, the diminutive architects of the delta, responsible for the formation of many islands, but their mounds also serve as biodiversity hotspots, supporting a large variety of bird, animal, and plant life.

main channels of the delta. These upper reaches of the Okavango, just beyond the permanent swampland, are key to the transformation and development of the vast alluvial fan, and to the annual progress of the flood. This is where "channel switching" first begins to take place – a complex and fascinating process that determines the shifts and excursions of the channel water, not only annually as the flood arrives, but also over a timescale of hundreds or even thousands of years. Beyond this geophysical switching process there are two less likely architects of the delta whose disparity in physical size belies their equal importance. One is the humble and tiny termite, the other the Okavango's portly and bellicose hippos. Both play a vital, yet slightly different role in the evolving shapes and patterns of the Okavango.

The few weeks while the flood seeps through the permanent swamp is a time of expectation and intense preparation for the film team. Our greatest challenge, the filming of the great flood, is soon to begin. I spend many hours in the plane, drifting over the thousands of islands and convoluted web of waterways, accumulating vital waypoints on the GPS, scanning the progress of the water and trying to discern a mental picture of where, and when, the flood will take its unpredictable course this year. Knowing these details is critical, as it will determine every facet of our planning and logistical operation for the next few months. Yet despite having worked here for over 20 years, I still find it nearly impossible to predict exactly how, and when, and where, the complex flows of the delta flood will take shape. Many hours are spent consulting with "old timers" and safari guides, hydrology experts and biologists. Even more hours are whittled away around our campfire debating the flood's probable rate and course, though we all know from past experience that, however fastidious our preparation, the Okavango will inevitably take us by surprise.

Filming the oncoming flood tide

The camera vehicles, our trusty Land Cruisers, are readied for their toughest challenge – hours and days of wading in chest-deep water, through lilies and reeds and aquatic plants, crossing rivers and flooded grasslands, churning through cloying mud and forging through thickly wooded islands. Every detail is checked and rechecked; we fit breather hoses to axles and gearboxes, extend exhaust pipes above the waterline, insulate electrical components, test heavy-duty recovery winches, fit custom-made camera boxes and accessories, and waterproof as much as we can. We have built a new rig for this season – an 11m (36ft) crane, with a remotely controlled head for the camera, fitted onto a custom-designed Land Cruiser. We hope that this sophisticated addition to our equipment will provide new views and perspectives on the coming of the flood. The

Termite mounds (right): these distinctive features of the Okavango, some reaching 4m (13ft) high, are the foundation upon which many of the islands are formed. Clays brought by the termites from deep underground provide a rich substrate, and passing birds deposit seeds from pioneer trees that will begin the process.

"swamp truck", our highly modified and somewhat ancient beast of a machine specially rebuilt for swamp conditions, is rolled out and readied for action, its monster mud-tyres fitted and the thundering V8 diesel engine overhauled. Camping equipment is carefully packed, warm winter clothing stuffed into waterproof bags, and carefully selected food provisions boxed and loaded onto the vehicles.

The Okavango is an extremely difficult place to make wildlife films, and it requires great skill to keep a camera crew safely and productively in the field. The challenges of the terrain are complex and potentially hazardous, the locations wild and remote, the animals often difficult to find, the logistics unbelievably convoluted, and the physical conditions frequently hard on man and machine. Yet we love filming here for these very reasons. We spend most of the year as a small team living in the wilderness, camped under a star-spangled Kalahari sky, surrounded by wild animals, hardly ever seeing another human soul, and we wouldn't have it any other way.

Finally we are ready. As the last light curves towards the horizon, and the winter stars explode in their glorious millions across the crisp desert night, someone cracks open a beer and mutters, "Eat, drink, tomorrow we ride!" misquoting Isaiah, Omar Khayyam and Gandalf all in one phrase. It is time for us to go and meet the oncoming wave.

It's a punishing two-day drive along winding sandy tracks deep into the heart of the delta, our vehicles straining under nearly a ton of equipment. Rigging our basic rustic camp on a small island, amid a shady

Filmmaking in the flood (above): our location presents unique challenges for man and machine.

The new flood (right): a spectacle from the air as water spreads across the floodplains and surrounds thousands of emerald islands.

cluster of motsaudi and mokuchum trees, we look out over the dry, expectant floodplains stretching around us on every side. With luck, the flood will arrive here within a few days. Around the campfire, enjoying a simple pot-roasted meal and steaming fresh bread made in the coals, the conversation turns to stories of our many years of filming in the bush. The time a hyaena raced in and grabbed our dinner from under our noses; the fearful night when our tents were repeatedly charged by a herd of 40 screaming, furious elephants; the evening a beautiful male leopard strolled nonchalantly into camp only yards from us and stole someone's shoes; the many, many times we have had vehicles bogged down for days in the swamp, their soggy interiors filling with startled fish; following lions at night through miles of water, soaked to the bone in near freezing conditions; climbing into our tents to find the bed already occupied by a 3m (10ft) cobra; young lions in camp stealing our soap and toothpaste; or the day we had to dive underwater a hundred breathless times to repair the clutch on a stricken vehicle that left us stranded, many days from help. These and many other stories roll on late into the night, and we crawl thankfully into bed-rolls, our heads filled with anticipation of the challenges that morning may bring.

Paradise regained

The next day, after rigging all the camera equipment onto the vehicle, we set off to explore. Driving up the dusty floodplain, it's clear the land has dried to extremes. The grasses have withered to dust, and the last pools in the small channels are nothing but cracked mud. Catfish skulls lie scattered on the baked clay, victims of fish eagles and marabou storks, who would have feasted on this rich harvest as the catfish pool dried, stranding its occupants and committing them to a certain fate. For them, the arrival of the flood is a few weeks too late. We drive for several miles along the dry channel, and then, rounding a corner, suddenly we see

Growth of the islands

One of the ways an island in the Okavango can develop is by the construction of a termite mound. Seeds deposited by birds or mammals take root in the mound and grow into a small woodland, marking the beginning of a tree island.

As trees and grass grow on it, animals come in to feed and deposit droppings that combine with fallen leaves to increase the volume and nutrient content of the soils. The vegetation is also continually capturing carbon and other nutrients from the atmosphere and adding these to the soil in the form of decaying leaves.

Slowly, over time, more and more topsoil is deposited and more and more trees grow, eventually forming a tree island. Water is continually sucked up through the roots of the vegetation, whilst the dissolved nutrients are not all absorbed, leading to a build-up of nutrients. This can cause a strong halocline – a separate salty layer – between the saltier ground water in the centre of an island and the surrounding fresh water. The continual transpiration (emission of water by the plants) causes higher and higher concentrations of salts to accumulate in the middle of the island, eventually resulting in the trees and grasses dying off in succession, depending on their salt tolerance. Ultimately some islands end up with a salt-crusted white patch in their middle that is totally devoid of vegetation.

The constant state of change that characterizes the Okavango prevents the area being poisoned by excessive salts. One of the key drivers of this dynamic system is, of course, the highly variable annual flooding, which not only fluctuates drastically in volume year by year, but also differs in distribution, with floodwaters being sometimes more concentrated in the east, and sometimes in the west, due to blocked channels or tectonic shifts.

dozens of egrets, storks, plovers, lilac-breasted rollers, and starlings, all feeding excitedly in the channel bed. We know exactly what these signs mean. Immediately we prepare the camera rig, knowing that every second counts.

Minutes later, we are hunkered beside the camera hastily set down in the dusty remnants of the ancient riverbed. As we ready the equipment the sand shifts minutely underfoot, tumbling into a mosaic of small fissures and cracks in the rock-hard mud. The cruel midday heat compresses an already stifling blanket of silence. Across the parched plains to the west a dust devil, twisting upwards in slow motion, whirls a dark funnel of dust hundreds of feet into the air, vultures riding on the updraft. Sand grains blow among a few desiccated stems of grass. On the sand ridge of the riverbank a few feet away, a scarab beetle staggers with desperate, slow-motion movement through the heat haze, step after laborious step. The waiting seems interminable. A ripple of white-hot wind tumbles a dried out insect exoskeleton across the shifting sands of the riverbed, and the eyeless, skeletal gaze of a long-dead catfish skull stares from the sand.

We, and the thirsty land, have been waiting for this moment for six months.

An hour later, our muscles cramping with fatigue in the searing heat and the camera almost too hot to touch, we hear the faintest sound of water, a burble of trickling liquid. Slowly, almost imperceptibly, a dark stain appears on a small ridge of sand 20m (65ft) upstream, creeping slowly down the face of the shallow sand slope. The tiny trickle becomes a rivulet, a snaking, mercurial thread of precious liquid that squirms towards the camera, throwing up miniature clouds of dust ahead of the oncoming meniscus. Within moments the thread becomes a pulsing tongue of moving water that swells and gently spreads, surging towards the camera. I frantically shoot, capturing the ephemeral moment of the event in a series of rapidly snatched extreme close-ups. Within seconds the water is all

Okavango islands (above): an estimated 15,000 islands are dotted across the Okavango. African ebony trees, palms, fig trees, and many other plant species provide sanctuary for a profusion of animals.

Papyrus islands (above): this suddenly abundant vegetation provides important roosts and nesting places for a variety of birds in the larger lagoons.

Wetland residents (above): more than 100 species of amphibians and reptiles live in the Okavango, a testament to its importance as a habitat.

around us, sliding past a few inches under the camera lens, filling my screen with a sheet of glistening, undulating water. The flood has arrived.

Suddenly the moment is filled with the sound of screeching, cackling birds. The air is alive with a stabbing, pecking, swallowing, flapping maelstrom of activity, as white egrets, yellow-billed storks, blacksmith plovers, and spectacular lilac-breasted rollers snap up insects at the head of the flood. We grab the camera and move quickly downstream, looking for a good vantage from which to film this spectacle. As the water surges through a narrow gully, we come across a pitiful sight. Flushed from its burrow, a small field mouse struggles in the water, and scrabbles up the half-submerged grass stems, desperately seeking refuge. But its respite is brief. A black-shouldered kite swoops in, too fast for us to ready the camera, and snatches the bedraggled rodent. Escape from drowning is almost inevitably followed by death from aerial assault, as the Okavango's raptors relish this season of plenty.

The miracle of the flood

And so, in June, as the floodplains reach their driest winter condition, the miracle of the Okavango unfolds. The flood disseminates slowly through these vast tracts of land, often moving less than 1km (½ mile) a day. Spreading gently into the dry grasslands, rejuvenating small channels, slowly refilling the shallow pools, surrounding small islands, the waters move inexorably southward. Occasionally the flood startles us with its sheer rapidity: camped one evening by a bone-dry lagoon, we wake in the morning to find it completely full and teeming with birds – an overnight transformation that is as magical as it is unexpected. It will be months before the slow inexorable tide reaches its southern limits. When it does, 12,000sq km (4500sq miles) of dry Okavango delta will have been transformed into a water wonderland – a verdant paradise in the middle of the Kalahari Desert.

A few weeks after filming the first flood arrival scenes, our small fly-camp is completely surrounded by water. Overnight the silent surge has crept down the floodplain and around us, its midnight arrival betrayed by the sudden ear-splitting chorus of thousands of reed frogs and bell frogs. We prepare for a major expedition to the north-west, a journey that will take us across several rivers and many miles of flooded grasslands. A recent recce flight revealed some exciting animal gatherings, and we are itching to get going. But preparedness is crucial, and we carefully ready the vehicles for a major swamp mission. On a map, the distance is less than 20km (12 miles), but we know that it could easily take two days.

A low golden mist hangs over the water at dawn as we set off in the swamp truck and a back-up vehicle. Immediately we are into the swamp, the waterline climbing steadily up the vehicle sides until it's over the bonnet and pouring into the cab through the windows. We forge slowly forward, pushing a bow wave, the surge ahead of us parting the lilies and reeds, and startling small fish that leap clear over the bonnet. Already we are soaked – water swirls over the gear lever and the seats, we drive with our legs entirely submerged, and the chill winter wind gnaws at our bones and nerves. Occasionally the vehicle's tyres lose traction, and we carefully reverse and manoeuvre until we have forward motion again. It's a slow and nerve-wracking process, not for the faint of heart.

A small error of judgement suddenly plunges the lead vehicle into a submerged ant-bear hole and we become seriously stuck. We dig underwater, shift mountains of mud, raise the vehicle by 1m (3ft) on high-lift jacks, wedge sand ladders and old logs under the tyres, and use the

Saddle-billed stork (left): these colourful birds join the profusion of bird species that hunt small fish, amphibians, and displaced insects at the head of the flood.

A time of plenty (next page): at the height of the flood, thousands of elephants enjoy the water and rich grazing. During the flood season they switch their diet from mainly browsing to grazing, and spend hours a day cropping the rich grasses nourished by the floodwaters.

recovery winch from the back-up car, all in the icy wind and up to our elbows in crocodile-infested waters. The team is well versed in this kind of routine recovery, but it is three hours before we are on our way again, and even longer before we are warm. But spirits are high and the swamp looks glorious, a breathtaking tapestry of blues and greens, and nature's splendour is all around us. Fish eagles and storks arc in the sky above; small groups of lechwe canter through the shallows, casting sparkling nets of backlit spray; we come across giraffe, kudu, waterbuck, zebra, reedbuck, and wildebeest moving belly deep through the shallows. All manner of water birds fill the air – thousands of white-faced whistling ducks wheel in noisy flocks, saddle-billed storks, herons, egrets, plovers, and small groups of the gorgeous pygmy goose are everywhere we look. Truly, this is paradise.

Lechwe
Kobus lechwe

These highly adapted swamp antelope have evolved to be quite at home in areas such as the Okavango, one of the few places in southern Africa where they can be found in significant numbers. The population in the Okavango region is estimated at around 40,000, making them one of the most abundant large mammals here.

Lechwe are not particularly fast on land and are often taken by lions, crocodiles, wild dogs, and sometimes leopards. Despite this vulnerability, they have evolved splayed hooves to keep them from sinking into the mud, and their raised powerful hindquarters enable them to outrun predators in mud or shallow water. During the dry months, the lechwe retreat towards the permanent swamp, but as the annual flood arrives, they spread with it, grazing on the tops of new grasses that emerge from the floodwater. They spend their days belly deep in the water, safe from predators, and rest up on small isolated islands at night.

KEY FACTS

Height	90–100cm (3ft–3ft 3in) at the shoulder.
Weight	70–120kg (150–265lb)
Horns	males have powerful lyre-shaped horns and will sometimes fight to the death with other males who invade their territory.

Fields of green (right): the flood turns the Okavango's floodplains into perfect grazing for the delta's 40,000 water-loving red lechwe.

Dealing with the water 1 (above): all the plains animals have to cope with the watery environment when the flood arrives. Here a magnificent male leopard leaps a deep channel as he moves from island to island.

A profusion of life

Two gruelling days later, cold and tired and covered in mud, we reach our objective and pitch our tiny fly-tents in a spectacular area of newly flooded grasslands and tree-studded islands. The air is alive with mosquitoes and we slap and curse as we work, being extremely vigilant in the bush because we know that the rising water flushes deadly cobras, puff adders, and mambas from their floodplain hiding places and onto the islands. The flood is moving in on several fronts all around us, and immediately we ready the camera for action. That evening after dusk as the flood moves into a nearby plain, the air is thick with millions of emerging midges, a cloud of insects miles wide, hanging above the grassland like rolling mist. Unexpectedly the sky suddenly fills with tens of thousands of pratincoles – swift, agile birds that streak back and forth through the insects in a spectacular, high-speed feeding frenzy that none of us has witnessed before. Sadly there's insufficient light to film, and within half an hour the birds, and the crepuscular light, are gone.

We spend a few weeks here in the upper delta, driving out an hour before dawn and back to camp well after sunset, filming dozens of scenes – privileged glimpses of new life in the delta as the magic of the Okavango flood unfolds. As the broad front of water advances, only inches deep across the floodplains, insects are forced from hiding places and thousands of birds feed on this flush of nourishment. Pairs of elegant wattled cranes dance a glorious mating ballet; thousands of open-billed storks wheel in the clear sky and descend en masse to feed on snails; small rodents swim for safety; and a deadly puff adder winds its way boldly past our vehicle and across the water to a nearby island. The flood rushes down hippo paths and channels – golden, clear water that spawns a million glittering bubbles in its wake. Within hours at the shallowest edges, the water is teeming with billions of zooplankton, hatched from tiny eggs that have lain dormant in

Feeding frenzy (left): Yellow-billed storks at the headwaters of the flood.

Dealing with the water 2 (next page): a common sight during the flood season, herds of impala race through the water.

Mokolwane palms – an elephant bull's favourite

These towering palm trees (*Hyphaene petersiana*), so distinctive of the delta, are a delicacy for elephant bulls, which will travel hundreds of kilometres to feast on their gingery fruit.

A bull places his trunk up the thick bole of the 20m (65ft) tree, or leans his forehead against it, and then uses his immense bulk and strength to give it a good shake. Each bull has learned from his elders how to push at a particular rate or frequency, setting up a resonance in the tree that violently shakes the top, dislodging the hard fruit. These then rain down upon the expectant animal, who picks them up one by one from the ground. Experienced bulls appear to be able to distinguish male from female trees, and carefully select the female trees that carry an abundance of fruit. The process benefits the trees, too: the seeds enjoy increased germination chances once they have passed through the gut of an elephant.

Baboons are also fond of the tasty pith surrounding the palm fruit, although they are unable to eat the central seed, a very hard kernel often referred to as "vegetable ivory" due to its polished whitish appearance. Inhabitants of the Okavango use the sap of mokolwane to make palm wine, and weave baskets out of the leaves of young trees.

A special technique (above)**:** Bull elephants shake the palms with their trunks to dislodge the fruit. Younger bulls often don't have the weight or strength and will surreptitiously steal fruits dislodged by their elders.

the sand. Thousands upon thousands of fish move into the grasslands, the sun-warmed waters providing an ideal nursery for their offspring, who will feed on the flush of zooplankton and on minute organic particles washed down with the flood. Hippos bellow in the newly filled lagoons.

On the islands, we film one of our favourite Okavango winter scenes – great bull elephants shaking the tall mokolwane palm trees, dislodging tennis-ball sized fruits that rain down on them and are eagerly savoured for their gingery taste. Breeding herds of elephant are here too, after their long trek from the dry mopane country, and they plunge with gleeful delight into the newly filled channels, both adults and youngsters romping and splashing and revelling in the new water. I am always captivated by this scene and have, over the years, burned many thousands of feet of film as baby elephants frolic and tumble in the water, thwacking their little hosepipe trunks in delight, or squeal with frustration as they inadvertently stand on their own unruly proboscis, over which they have no control.

Adapting to the water

Finally, after days of searching, we find one of the highlights for which we came. Out of the early morning mist, a vast dark mass begins to take shape and suddenly, as the steam and dust and vapours clear, we see them – a thousand buffalo, bulky forms wading slowly through the shallows, their gentle bellows, pungent smell, and steamy breath filling the air, a skyline of gracefully curved horns, and anklets of glittering water sparkling around them with each steady step. Our excitement grows. We know that where there are buffalo, there will be lions – the "swamp cats" of the Okavango. Now our mission is clear – we need to stay with the herd for the coming days and nights, in the hope that the elusive cats will not be far behind.

Most of the large African mammal species have learned to adapt in some way to the swamp conditions, and each uses a unique strategy to cope with the challenges. Every year, these massive herds of buffalo, sometimes 3000 strong, thunder into the delta to graze the sweet new grasses nourished by the flood. For the lion prides that follow them, there are two choices – starve or swim. Wading through miles of chest-high water, swimming deep channels, they hunt the great buffalo herds. Growing up in the swamp, the lion cubs learn to swim early in life, and the females in the pride regularly lead the reluctant youngsters through the shallows to acclimatize them to this distasteful medium. Crocs present a genuine hazard for the swamp cats, and the experienced members of a pride have learned to check carefully for crocs whenever they cross deep water. Hunting shoulder to shoulder with the cats, hyaenas have also adapted to swamp life. They trail and sometimes rob the pride, and may stash meat in the shallow water for leaner times. Often the roles are reversed, with the lions plundering the hyaenas' kills. Nowhere else in Africa have these great cats learned to specialize to the degree that the swamp cats have done. We follow the buffalo for two days and night, scanning all the while for signs of lions. Finally, on the third morning, bleary-eyed and clutching steaming mugs of coffee, we hear a faint splashing on a nearby floodplain, barely audible above the grunting and heaving of the buffalo. We creep through

Drinking their fill (left above and below): as soon as the incoming flood fills the deeper lagoons, the elephant herds arrive. They plunge into the water with delight and often play for hours at a time. Young elephants even wrestle and climb on each other's backs.

Buffalo bonanza (next page): thousands of buffalo move into the delta to enjoy the rich grasses. Smaller groups merge together, and herds of 2000 are not uncommon in certain areas.

the bush for a better vantage. It's a spectacular sight: a huge dark-maned lion is striding belly deep through the swamp towards us.

A hundred metres (330ft) to one side comes the whole pride – four older lionesses, five adolescents, and three small cubs. The little ones are almost completely submerged, swimming frantically to keep up with their mother, their tiny noses wrinkled and snarling at the discomfort of the cold water. We quickly move back to the camera car and quietly manoeuvre into position to film. The lions continue to swim towards us, unaware of our presence, and we are rewarded with some amazing shots. Over our many years in the Okavango we have spent long periods struggling to film these swamp cats, and everyone is delighted that we have captured these brief scenes. The lions move nonchalantly towards the buffalo herd, then settle down to dry off in the winter sun. We know from past experience that filming them hunting these buffalo will be extremely taxing. We have only a few days for the attempt; then, we need to move on to cover other aspects of the flood.

Swamp cats

In the seasonally flooded areas of the Okavango, there are several prides of lions that have adapted to life in the swamp. For half the year, they live among the dry islands and floodplains. But when the flood arrives, a large proportion of their home range goes underwater, forcing the cats to wade or swim between islands.

Recent research has shown that, unlike their savannah-living cousins, these "swamp cats" have also adapted their territorial behaviour, and that during the flood season, when the available dry land area is dramatically reduced, the prides' home ranges may actually overlap by up to 60 per cent – a remarkable change in lion social behaviour brought about by the flood. In addition, it appears that "swamp cat" pride males spend much less time patrolling their territories than those in dry areas, with the result that they spend more time with the pride and less time in conflict with neighbouring pride males.

Filming paradise from the air

At the height of the middle delta flood in June, back in our small camp in the delta, the morning stillness is ruffled by the insistent throb of a helicopter. The production office in Maun has spent months preparing for a week of aerial filming – organizing a 7-tonne 4x4 truck to bring in hundreds of litres of chopper fuel and additional camp kit, re-supplying us with food and vehicle parts, co-ordinating specialist equipment, planning radio and satellite communications, and a dozen other logistical details.

Peter, a long-time friend, superb chopper pilot, and old delta hand, slides the helicopter neatly down onto the grassy edge of the island. The chopper is fitted with a Cineflex heligimble mount – a remarkable piece of high technology equipment that allows us to shoot rock-steady images on a long telephoto lens from the air. This is significant, as we can remain far enough away from the animals that we cause no disturbance and can capture their natural behaviour. We have to plan our airborne time carefully, as distances are great, and the helicopter is expensive and has limited endurance. To avoid wasting helicopter time I will fly early morning sorties in the Cessna, using it as a spotter plane, directing the helicopter from the air to scenes and locations we want to film.

We spend several clear winter mornings flying over the spectacular mosaic of the flood, filming animals in the water, hippos surging down crystal-clear channels, and the floodwaters inundating new dry areas. From this privileged vantage, the sheer scale and beauty of the Okavango is breathtaking. For those of us that know the delta well, there's always a fascination with being airborne over the swamp, as we spot new areas of flood, changes in the channels, subtle shifts in animal concentrations from year to year. As we climb high over the central delta on the last morning, the furthest reaches of the Okavango come into view, 50km (30 miles) away in the crisp winter air. These distant floodplains are still dry, and it could be weeks before the water reaches them, if indeed it makes it that far at all.

Fire and flood on the drying plains

A few weeks later, we are back at Gomoti camp. It's late in July, and the flood has still not arrived. Scanning the area from the plane, I can see that the flood is not far away, but how fast it is moving is difficult to gauge from

Swamp cats (above): lion prides in the central areas of the delta have adapted to life in the swamp, and regularly wade or swim the deeper channels.

Complex logistics (above): an aerial supply drop to the film crew in a remote area of the delta.

the air. Sweeping over Hippo Island, I can see that the hippos' plight has deepened. There is very little water in the pool, and the scant grazing that remains is poor and distant from their muddy refuge. But things are about to get worse – from the plane I spot a small smudge of smoke, the faintest hint of fire, not far from the island.

I immediately land, and we race out in the camera vehicle. The wind is up, and the faint billow of smoke has turned into an inferno in a few minutes. Flames are tearing at breakneck speed through a long island, the small palms igniting into fireballs with astonishing violence; tongues of fire leap 18m (60ft) into the air and catch the fronds of the tallest palms. Choking white smoke boils all around us as we film, racing the flames in an attempt to capture close-up shots without being incinerated ourselves. At one point the wind switches, and the fire surrounds us – we are forced to burst through the fire-line in the vehicle to escape, and continue filming. Hundreds of birds – kites, rollers, drongos, hornbills, and swifts – fill the air, swooping through the smoke to catch insects dislodged by the fire. A column of white and slate-grey smoke rises thousands of feet into the cobalt sky.

By evening the wind has died, the birds are satiated from their feast and, careful not to tread on puff adders displaced by the fire, we survey the damage. Hundreds of hectares have burned, and eddies of smoke drift up from the smouldering remains of old tree stumps. For the hippos this is a final cruel blow. Most of their nearby grazing has gone up in smoke. Occasional fires like these are good for the Okavango, and many thousands of square kilometres burn every year, releasing valuable nutrients to the soil. The bigger fires are usually fanned by strong winds, driving the flames rapidly, creating a "cold" fire, which does little damage. Many of the delta's trees and plants are superbly adapted to fire, and larger animals simply move away, so the only real victims are the smallest mammals and insects who, if not consumed by the flames, are taken by birds as they scrabble for safety. The real conservation concern regarding fires are those areas burned repeatedly at the hand of man – an annual repetition that ultimately, if not controlled, could turn the delta into a dusty wasteland.

Fire alarm (above): dramatic bush fires occasionally rage across the Okavango. While many of the delta's plants are well adapted to fire and only a very few unfortunate small mammals and insects are unable to flee the flames, frequent repeated burning is a genuine threat to the delta ecosystem.

A few days after the fire, we discover another remarkable natural event that is about to develop. A large lagoon, set at the end of a subsidiary channel, has dried from its previous glory of 500m (⅓ mile) wide into a small muddy waterhole, barely 3m (10ft) across. What catches our attention from afar is the glimpse of dozens of noisy fish eagles that suddenly burst out of the surrounding trees overlooking the waterhole. What has attracted them from miles around, and now enthrals us as we investigate, is the boiling, tumbling maelstrom of hundreds of catfish, desperate prisoners in the rapidly drying sludge of liquid mud. We mobilize quickly, driving back to Gomoti camp, gathering our fly-camping gear, and hastily rigging our camera crane. That evening we set up a camera hide next to the pool – I have a hunch that the fish eagles will be back in force and we need to be ready. I sleep fitfully in the tiny hide overnight, crumpled up amongst the camera gear, so there will be no visible movement early in the morning. At dawn, the air begins to resound with the glorious calls of the eagles as they gather for the feast.

A marvellous tapestry of colour and texture (next page): the Okavango from the air. A small group of elephants moves through drying delta grasslands, creating an intricate pattern of trails across the landscape.

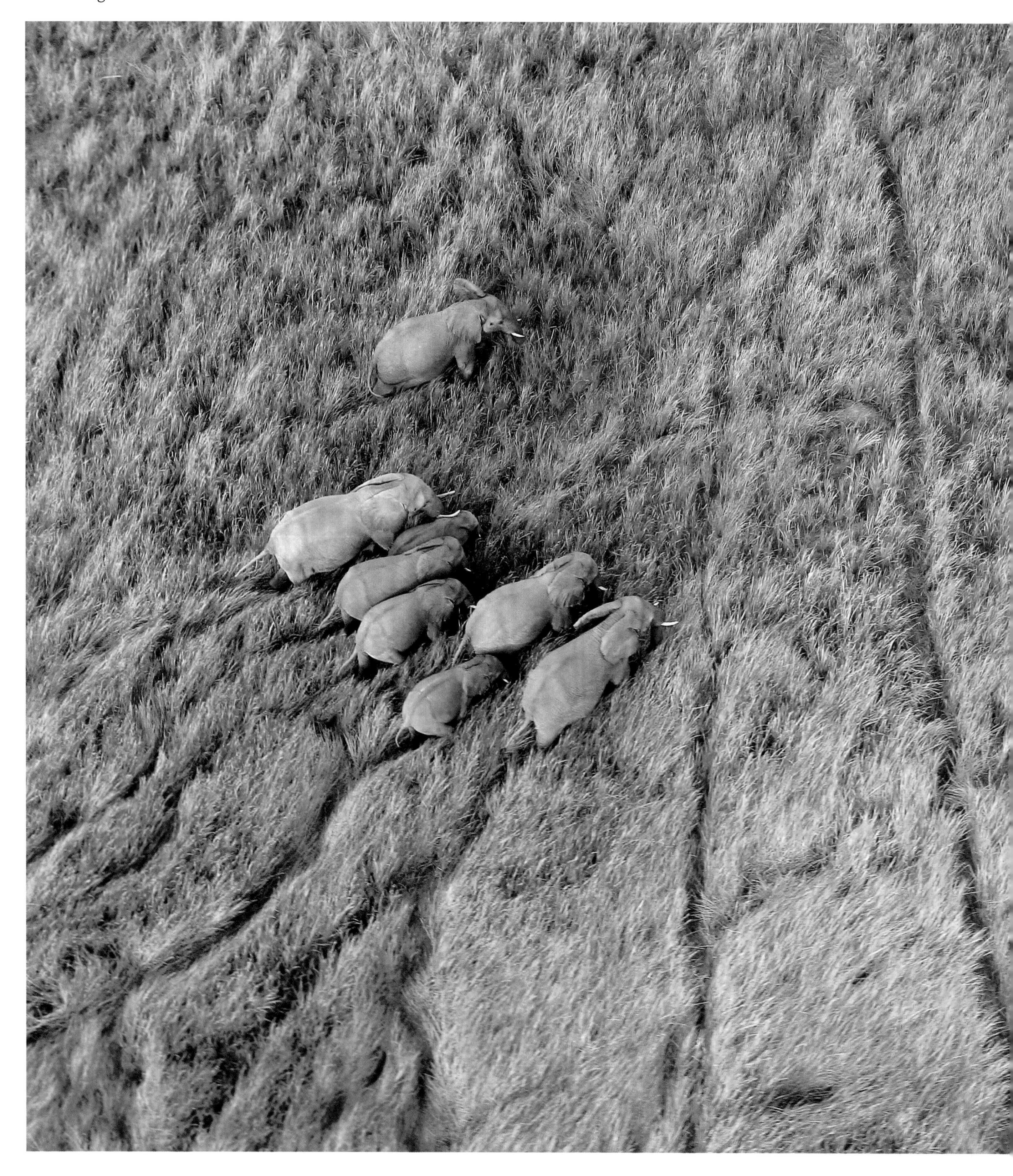

The sun begins to filter through the trees and I film the eagles as they arrive, one by one, swooping in on giant wings to the pool. For the next hour there is a free for all. Eagles hop into the edges of the pond, grab hapless catfish and drag them thrashing and squirming to the bank. The fish go crazy at each attack, a heavy mass of flapping, writhing, slippery bodies, fins and tails whipping the muddy, chocolate-mousse sludge. The eagles squabble amongst each other, abandoning all normal dignity and grace in a frenzy of shrieks, swiping talons, and flapping wings.

For three days the early morning assault continues. The pool is drying rapidly and the fish are dying for lack of oxygen. We film their desperate plight with the camera crane. But what we know, and are waiting for, is that their reprieve is imminent. On the fourth morning, as the remaining fish have lapsed into complete apathy and appear to have given up, their little miracle happens. The incoming flood breaches a sandbank, rushes headlong down a hippo path and surges into the pool. In a matter of seconds the whole pond erupts in a flurry of splashing excitement, the desperate fish gulping at the fresh oxygenated water they so desperately need. Within minutes the pool is 1m (3ft) deep, the fish submerge, and the fish eagles take to the air, soaring gracefully up on the thermals of the morning.

Back at Hippo Island a few days later, the miracle repeats itself on a grander scale. Fresh clear floodwater surges dramatically into the hippos' muddy pool, bringing new life and deliverance from the foetid slime of the desiccated lagoon. Once again, the Okavango flood has played out this miracle of salvation, and for another year, these hippos can pass their days in the cooling waters, feed on the rich grasslands that the flood will stimulate, and languish, fat and happy, on the sandy white banks of their luxuriant channel.

For those that know the Okavango well, this is its most dazzling phase. The temperament of the inundated floodplains and islands shifts almost imperceptibly every day, and we delight in a thousand tiny details as animals and birds feed and breed and move with the flood. Everywhere there are new enchantments – little moments of animal behaviour never seen before; the spectacular tangerine pre-dawn light reflected on silver waters; skeins of ducks whistling through the chill morning air. Of course there are equal spectacles with large mammals and Africa's charismatic big predators. But somehow it's the small details, the infinitesimal, subtle shifts of life or light or breeze, that always make us pause and savour our privileged existence in this vast, unspoiled wilderness.

Fig trees – islands of life

One tree, hundreds of animals – many creatures rely on the Okavango's spectacular fig trees for survival. Each species usually has its own associated species of fig wasp for pollination.

The large sycamore fig (*Ficus sycomorus*), is found growing near water, forming an integral part of the riverine forest, and also occurs in mixed woodlands. This is one of the most spectacular trees in the Okavango, achieving heights of up to 20m (65ft) with a very sturdy trunk whose flattened buttresses prevent elephants from pushing it over. Its light and soft wood is used by local peoples to make drums, grinding mortars, and sometimes wooden dugout canoes called mokoros.

Fire is said to be generated by rubbing the twigs to create friction. Various reptiles, rodents, and sometimes even warthogs make their homes in crevices and holes formed by the convoluted trunks and roots. The tasty figs are also a vital source of food for myriad creatures ranging from tiny insects, small fruit bats, baboons, and monkeys all the way up to the majestic elephant.

The majestic fig tree (right): one of the most spectacular trees in the Okavango.

"Pirates" of the Okavango (left): Fish eagles are an iconic and charismatic feature of the floodplains, preying not only on fish but also on small mammals, reptiles, and birds.

Thousands of square kilometres of the delta are now, in the truest sense of the word, a swamp. Vast areas of grasslands are under water. A great network of waterways spreads over the land, dotted with thousands of islands, and linked by a complex of shallow pathways and deeper crossing points created by animals moving from one area of dry land to the next. At the fringes of deep lagoons, pelicans, yellow-billed storks, open-bills, and marabous breed on low-lying stands of water-fig trees. Sitatunga emerge from the papyrus areas. The lechwe, now truly in their element, feed in the shallows, safe from the marauding of lions and wild dogs. For a few weeks, the Okavango is blanketed in fields of green, and the delta displays its spectacular palette of emerald, viridian, olive, sienna, and cerulean blue.

The miracle evaporates

Yet even as these expansive swathes of grass emerge, the shallow water is beginning to evaporate. Gradually, the remaining water begins to pulse off the floodplains and slide back into the deep channels. Millions of fish,

adults and new hatchlings, follow this movement, returning to the deep water. Soon they will begin to move 100km (60 miles) upstream, and as they reach the Panhandle weeks later, another remarkable event, the "catfish run", takes place. Hippos and crocodiles do the same, forsaking floodplains and lagoons for the safety of the channels. By the end of August, as the remaining trickle of floodwater reaches its furthest extent and meets the southernmost fault line, that fortuitous subterranean dam wall, most of the delta is already drying once again.

And so, in September, as the knob-thorn trees adorn their branches with a flush of yellow flowers in anticipation of November's rains, the cycle is complete. All too soon, our filming time is over. We have spent over 200 days filming across the Okavango, witnessing scenes of drying and death, flooding and salvation. And every scene, great and small, has reminded us of the magnificence and fragility of this spectacular delta, whose existence depends on those great rivers of life, and whose continual cycles will always be driven by one of Earth's greatest natural events – the Okavango flood.

The life-giving flood (above): the cycle of life in the Okavango depends almost entirely on the annual arrivals of this astoundng quantity of water.

The Great Feast
South-east Alaska

The layers of landscape along the northern Pacific coast of North America are a photographer's dream. Changeable weather patterns make for surprise skies, sometimes cloudy and grey, at other times vivid and full of colour. Beautifully misty August mornings quickly burn off with the summer sun. Seas can on some days be messy and choppy and on others like glass, with the sound of whale blows or the wistful call of a loon being carried great distances along the surface. Here, in summer, the waters are packed with life: the plankton has bloomed. And the story of the plankton is the story of a great feast.

As sunbeams filter down through the emerald summer sea they illuminate a blanket of green specks that drift in the water: plankton. Six months previously these same waters were clear and empty. Now they feel packed with life. This dramatic seasonal change in productivity – and the wildlife that it draws in – is not the cause of one of our Earth's greatest natural events; it really is a great event in itself. The focal point of the great feast is the coast of south-eastern Alaska and northern British Columbia. The area is impressive not just for its wildlife but also for the sheer beauty of its land and its seascapes. The convoluted coastline, with its patchwork of passages, islands, and channels backed by dramatic mountain vistas, is breathtaking.

South-east Alaska
United States

Composed of a narrow, 1000km (600 mile) long stretch of mountainous land sandwiched between Canada's western flank and the north-eastern Pacific Ocean, south-east Alaska is a land of true wilderness. Snow-capped mountains and huge glaciers give way to the largest tracts of virgin forest in the United States, providing a home to the highest densities of grizzly bears and bald eagles found anywhere in the world.

Over 1000 islands make up the Alexander Archipelago, where land and sea mix together to form nearly 16,000km (10,000 miles) of coastline, with intimately linked straits, bays, and sheer-sided fjords. The seas off this coast are the most productive on the planet, and provide an important home to a host of marine life.

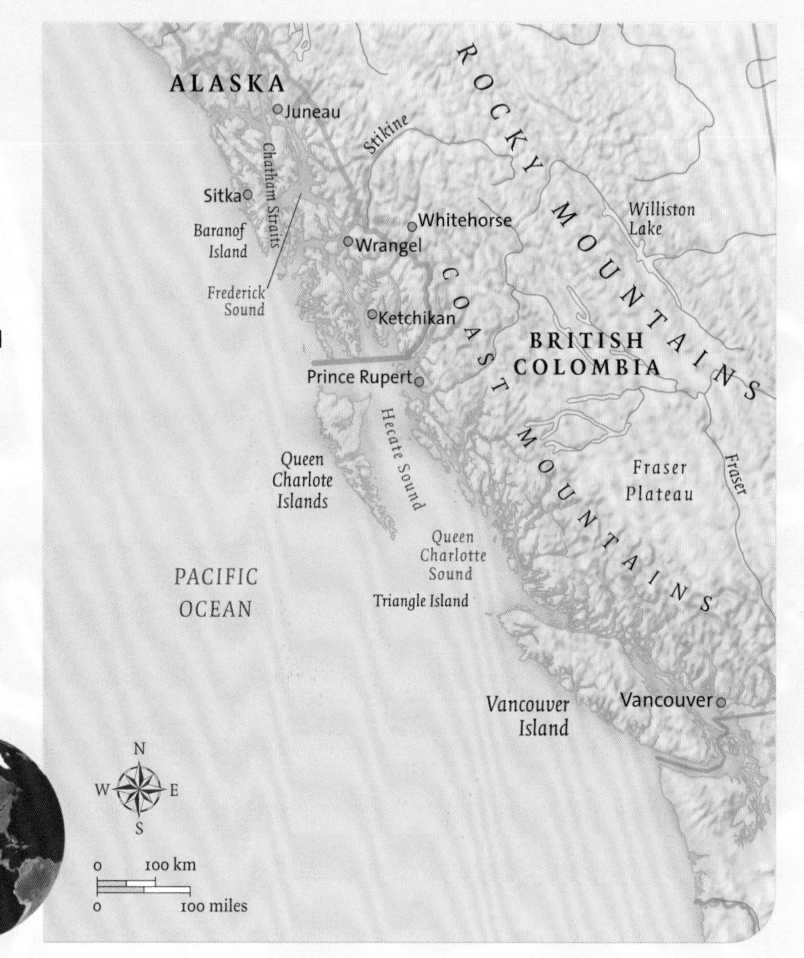

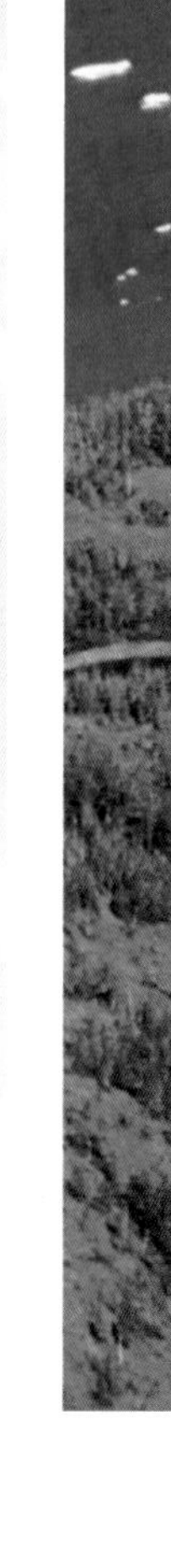

A patchwork of fjords and islands (right): the glaciers that carved out this twisted coastline are still active today.

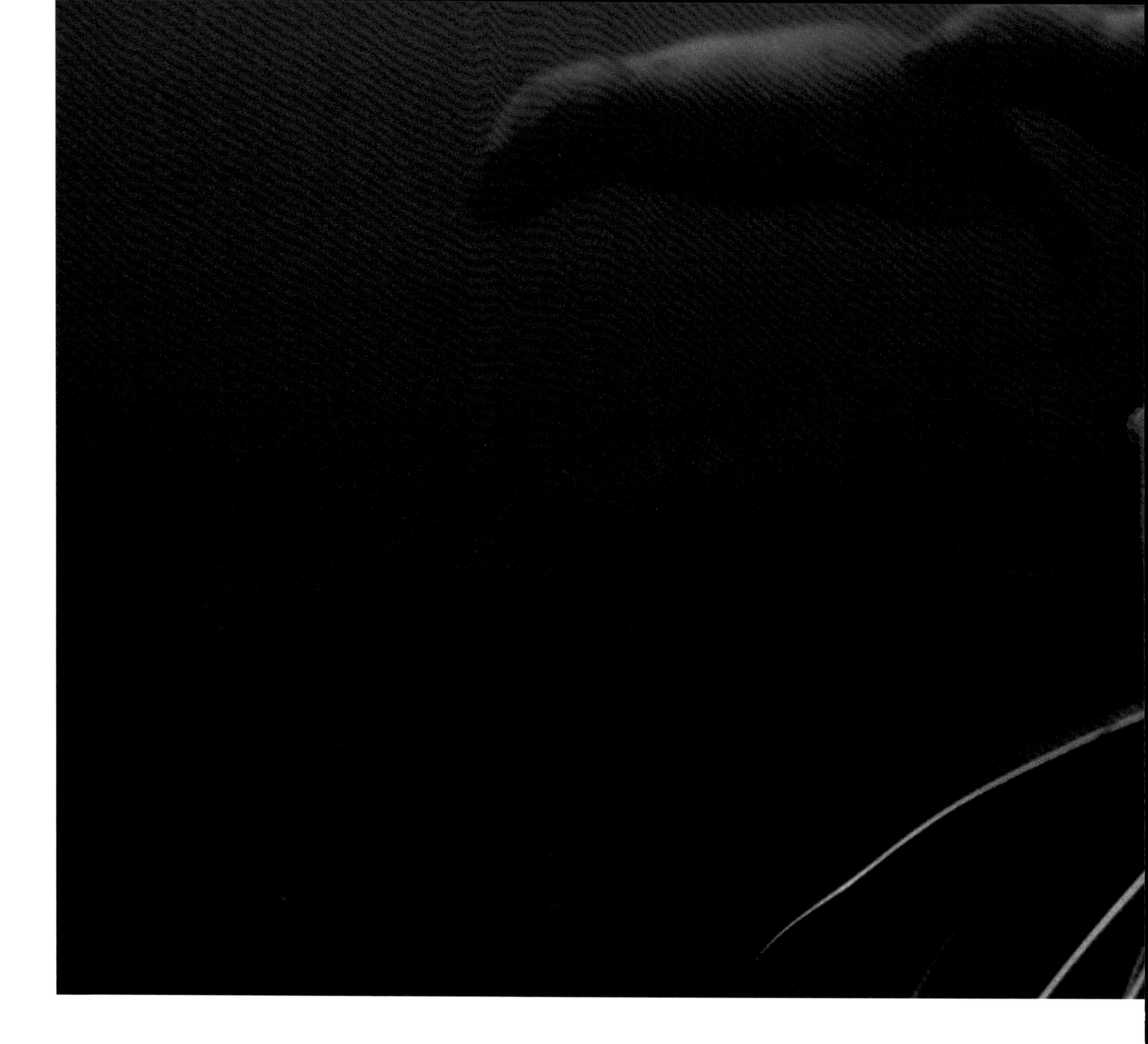

A pristine wilderness

This is a place to be admired but also respected in its wildness. Its grandeur is inspiring, but it reveals its secrets only with time and patience.

All over the world, wherever spring brings longer, warmer, and brighter days, phytoplankton blooms. But the conditions off this particular coast bring about a bloom like no other. This is no flower-filled, riot of colour – at first glance it appears nothing more than a green soup. But at its peak the body of plankton in these waters holds a greater biomass than that of the Amazon forests. And this richness supports huge shoals of fish that in turn feed a wealth of wildlife. Most astounding is that the plankton does not wither, die, and then decay on the seabed to decompose and release its nutrients back to the marine system. Rather it is almost all consumed, fuelling the entire food chain.

Though these waters are some of the most productive on the planet, they are not constantly so. The plankton bloom creates a seasonal change in the sea, the extent of which is not unlike that of the Pacific rainforests

that line the coast (as featured in The Great Salmon Run, see page 212). However, unlike what happens in these rainforests or on the great plains of the Serengeti, featured in The Great Migration (see page 108), where the climax of the event provides vistas crowded with wildebeest, the explosion of life here happens underwater and for the most part out of sight. But the evidence of the richness of the system can be seen in the animals that it draws in. These waters are home to sea lions, sea otters, porpoises, and killer whales, while humpback whales migrate from Hawaii – one of the longest migrations of any marine mammal. The world's highest concentrations of bald eagles are found here, alongside a variety of resident birds, while some migrants travel from as far afield as New Zealand. It is the different strategies that these wildlife characters adopt to get the most out of what the seas provide that allow us to understand a bit more about the complexity of the ecosystem here. And the real driving force for the wildlife is the abundance of fish that the plankton bloom supports, and the annual herring spawning, which produces literally trillions of eggs in the space of a few weeks.

The iconic Stellers (above): large eyes and dextrous whiskers help Steller sea lions find fish in the often gloomy waters. They will regularly forage at a depth of 20m (65ft) but have been recorded diving to over 300m (1000ft).

The bloom from space: the annual bloom is so vast it can be seen by satellite sensors orbiting the Earth. They show how January's lifeless seas **(above left)** transform to plankton-rich waters in summer **(above right)**. Imaging phytoplankton levels also shows the eddies and ocean currents that sweep the coastline.

Plankton

Plankton is generally divided into three broad levels:

Phytoplankton are tiny plants that are hugely important to the productivity of the seas, as they act as the primary producers of the oceans. This means that, like terrestrial plants, they are able to fix energy from the sun using photosynthesis, allowing them to produce organic matter from carbon dioxide.

Zooplankton are the first consumers of the food chain. They consist of small grazing animals that eat phytoplankton, and predatory animals that prey on small organisms such as other zooplankton and even small fish. Several species of larger animals such as fish and crustaceans also contribute to the zooplankton, as their early larval stages add to the planktonic soup. The zooplankton of south-east Alaska is hugely important to its ecology, as it in turn feeds the herring which fuel the entire ecosystem.

Bacterioplankton act as the recyclers of the plankton world, consuming and converting organic material produced by other organisms back into inorganic nutrients.

KEY FACTS

What is it?	any drifting organism that inhabits the oceans, seas, and bodies of fresh water.
Size	anything from as little as 0.1μm (1/10,000mm/1/25,000in) in the case of marine viruses, to 20cm (8in) and above in the case of organisms such as jellyfish.
Significance	plankton are some of the most important organisms on Earth, as they form the base of the majority of aquatic food chains.

To really grasp what drives the productivity of these waters on a planetary level, it's best first to see the sea on the small scale and look at the minutiae of life – plankton. The basic building blocks of life in the seas are phytoplankton, the small chlorophyll-packed particles which are the base of the aquatic food chain, and which bloom in the right conditions of sunlight and oxygen-rich waters. While they may individually be only 0.1mm (1/250in) across, the amount of chlorophyll collectively held in the plankton within a body of water can be so vast that it can be seen from satellite sensors orbiting the Earth. The bloom of plankton is in part driven by the current and tidal action along the coastline, which is swept by the Alaska Current, an eddy of the North Pacific Drift. This helps regulate the maritime climate, while the deep fjords and networks of islands ensure that the influence of this water is circulated all along the coast. In addition, the fast-moving Alaskan Coastal Current provides an important transition zone between the living communities in the shallows and those in the deeper water out to sea; it distributes plankton into the ideal living conditions of the sheltered inshore waters. These features also amplify the effects of the tides, creating strong tidal exchanges and producing areas of high currents. It is these currents stirring up the waters and bringing oxygen-rich water to the surface that fuel the phytoplankton bloom and produce concentrations of plankton to feed fish, birds, and mammals alike.

From the shore, the sheer magnitude of the life here is hard to grasp. Satellite images illustrate how productive these waters become in the summer months. But to see it close up you really have to get into the water, beneath the waves. At the beginning of the year the scene is quite different.

Phytoplankton (next page): the most basic element of the marine food chain. Without it, none of the larger animals would survive.

A glacial history and oceanography
Productivity of the region

South-east Alaska is a land of towering mountains, volcanic peaks, and steep-sided fjords that cut deep channels inland, causing the sea to penetrate deep into the landmass. This rich and varied topography is the product of a volcanic landscape moulded over thousands of years by periods of glaciation.

As little as 20,000 years ago, almost all of this now-productive region was covered by a layer of ice up to 1200–1500m (4000–5000ft) thick. Some 10,000 years later, the ice sheet began to retreat and gave way to a network of glaciers that carved their way through the landscape, creating U-shaped valleys with steep walls and level floors. The huge weight of these glaciers enabled them to gouge deep into the bedrock well below sea level and, as the glaciers began to recede, sea water flooded into the newly formed valleys to create the deep fjords that we see today. These fjords and the network of sheltered channels that run between the islands of the Alexander Archipelago amplify the effects of the tides.

In many of the world's temperate seas a spring phytoplankton bloom is initiated when light levels increase in response to warmer and brighter days. In these conditions, phytoplankton undergoes explosive growth, and a huge bloom is seen in the surface layers of the ocean. However, the stabilizing of the water column that occurs at this time restricts the vertical movement of nutrients such as nitrogen and phosphates that are vital to plankton growth. This can result in the dramatic slowing of the bloom in early to mid-summer, as nutrients are used up. In south-east Alaska, however, the increased tidal mixing that results from the unique topography can periodically bring nutrients up to the surface. This resupplying of the phytoplankton with the crucial ingredients they need for growth can often result in a prolonged bloom – meaning that more organic matter is directed into the food web, and the productivity of the region increases.

The oceanography also contributes to this area's amazing productivity. Hugging the coastline and looping through Prince William Sound, a fast-moving current known as the Alaskan Coastal Current or ACC provides an ecologically important transition zone between shallow, near-shore communities and the open ocean systems. Fed by coastal fresh-water run-off, it distributes plankton into the region's sheltered inner waters, where they have the ideal conditions to bloom.

A crucial link in the marine food chain (below): zooplankton. The bloom in phytoplankton feeds these tiny creatures, which in turn become food for the herrings and other fish.

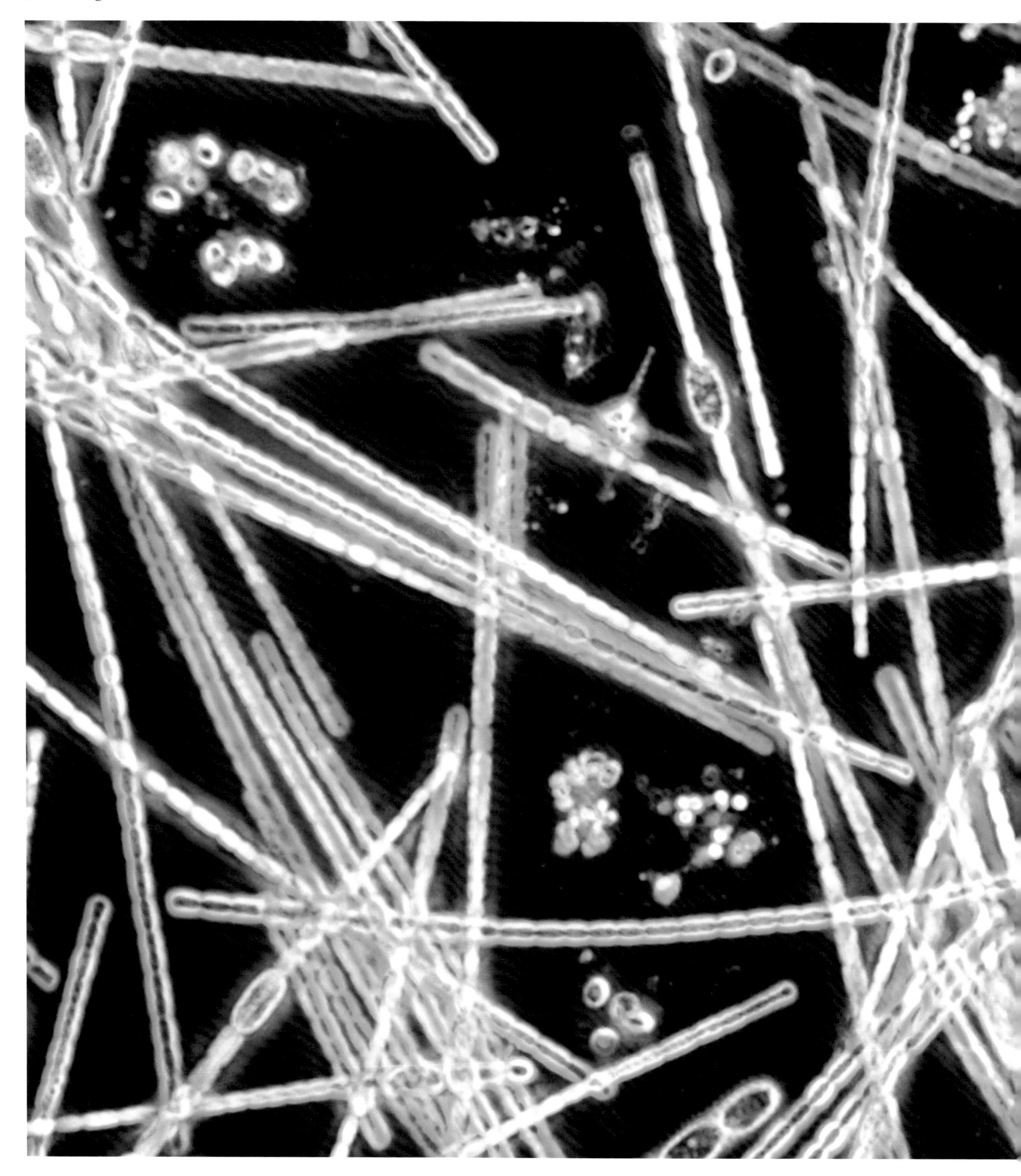

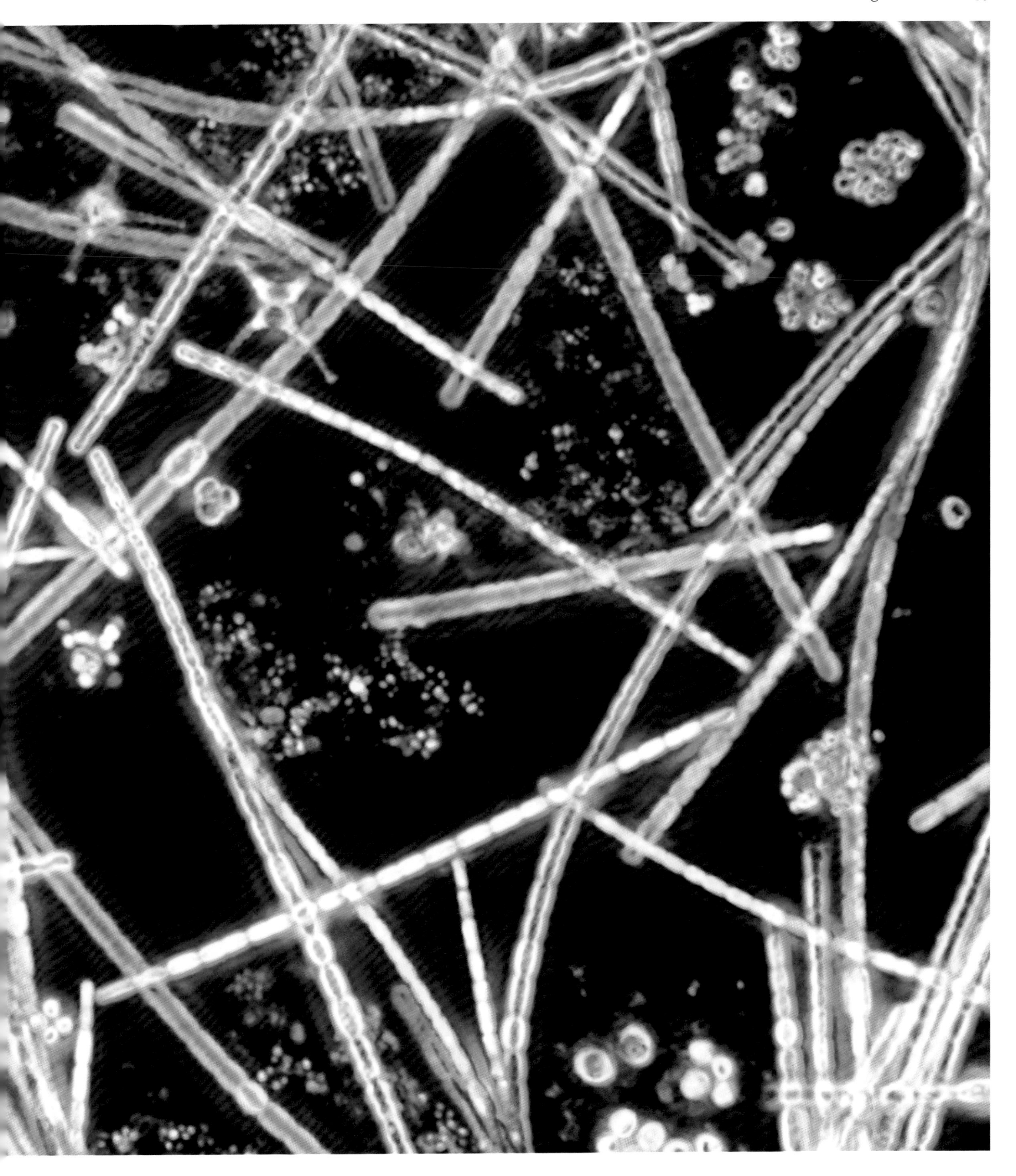

The richest seas on Earth
Joe Stevens

My first experience of diving in the Pacific North-West was in March. At this time of year the sea still feels dormant. The weak sun has yet to stimulate the first phytoplankton blooms of the year, so although the water is rich in oxygen it is clear and cold.

Winter

As I jumped from the boat, cold water hit my face, swept around inside my hood and seeped into my neoprene gloves. Above the surface the day was gloomy and overcast; underwater the seascape was clear, the water empty and the forests of bull kelp withered to stringy fronds that had survived being washed or broken from the rock during the winter. Through this world I floated with cameraman Shane Moore, looking for one of the region's most charismatic residents, the Steller sea lion.

Working in this cold water, a matter of degrees above freezing, it is the hands and face that shut down first, then slowly your falling body temperature turns your movements and even your thinking sluggish and laboured. After a number of hours in these lifeless waters there was little to distract me from the creeping cold feeling. Wearing drysuits and proper

An icy dive (above): Joe Stevens off the coast of Alaska in March. Cold water diving regulators and drysuits are the name of the game for anyone wanting to work in these waters.

Winter turns slowly into spring (left): the cold waters of south-east Alaska are rich in oxygen but require sunlight to fuel the first plankton blooms of the year.

insulation layers prevented us getting dangerously cold, but dive after dive, day after day, made me truly appreciate how tough the Steller sea lions that live here year round are.

For me it is the Steller sea lion more than any other animal that embodies the coastal Pacific North-West. For a start it is endemic to this coast, ranging from California all the way through the Gulf of Alaska. Its southern cousin, the California sea lion, is increasingly encroaching into the southern part of its range, and its characteristic "honk-honk" call (distinguished from the knarled growl of the Stellers) is heard more regularly in sea lion colonies up the British Columbian coast and even into south-east Alaska. But the Stellers hold court throughout these seas. They have opted for the rough-it-out approach to the harshness of life here,

and have the physiology and attitude to match. They are the brutes of the sea lion family, the largest and heaviest of their kind. An average bull can weigh in at 565kg (1245lb), the size of over five Mike Tysons, and have a skull larger than that of a grizzly bear with an equally impressive set of teeth. In a colony they appear boisterous and gruff, always either snoozing or bickering. As Shane aptly puts it, if there was only a handful of these animals left in the world they'd still crowd onto a small rock and complain about it. Nevertheless, their resilience epitomizes what's needed to survive year round in these waters. And so filming them meant we too had to experience some of their conditions.

For one of our camera teams, Warwick Sloss and Ed Charles, this meant camping near a colony in order to show how Stellers endure the

Pacific storms batter Triangle Island (above): this is Canada's largest Steller sea lion rookery. Stellers prefer breeding on remote, rocky coastlines like this, but have to cope with hazardous storms throughout the year.

Alaskan snowstorms in January. It got so bitterly cold that one morning the team emerged from their tents to find a dead bald eagle lying on the ground, fallen from its perch and frozen solid. The sea lions meanwhile lay shivering on the shoreline, clothed in a blanket of snow. They cope with this cold by having extremely thick blubber layers and high levels of endorphins. And unlike other sea lions species, the young can suckle for up to three years to help them make it through the harsh winters when other food is at a premium.

The way the adults manage to continue to find the scatterings of fish to feed on during the winter is truly remarkable, made even more impressive when you consider that they mostly feed at night. But with the rich seas of spring just around the corner, as the days lengthen the long lean winter comes to a very dramatic end.

Steller sea lion
Eumetopias jubatus

Despite being so at home in the water, Stellers are only semi-aquatic and need to come ashore to give birth and breed. The males arrive at their breeding rookeries on exposed rocks and beaches in early May, and fight to set up a territory – with particularly bloody battles occurring when opponents are evenly matched.

The females start to come ashore in mid- to late May and continue to arrive well into June. They give birth to a single pup a few days later after a nine-month gestation period, and remain with it constantly for a little over a week, after which time they begin to undertake short foraging trips to sea. Mating generally occurs shortly after giving birth, though the females delay implantation of their fertilized egg for 3–4 months, ensuring that the new pup is born during the abundant summer months. In good years they wean their pups gradually during the following spring, prior to the breeding season. However when food is less plentiful, pups take up to three years to be fully weaned.

KEY FACTS

Size	3.25m (10–11ft)
Weight	the largest of all sea lions, males weigh up to 1100kg (2500lb), twice the size of a grizzly bear; females 350kg (almost 800lb).
Food	a wide variety of fish, squid and octopus; also occasionally small sea mammals such as seal and sea otter pups.

Sea lion haul-out (left): a Steller pup suckles from its mother. Stellers are highly gregarious both on land and in the water, a behaviour that gives them protection from land predators and killer whales alike.

Seabirds

Seabird flocks feeding on hapless shoals of herring may look like a frenzied rabble, but there is an order to the chaos. When mixed-species flocks assemble like this, the birds can be separated into four distinct groups:

Initiators such as kittiwakes tend to instigate the gathering of a feeding flock. Other species in the area will watch these birds carefully and, as soon as the initiators begin feeding, they will move in for their piece of the action.

Divers are the real engineers of the bait ball. Most divers belong to the Alcids, a group that includes guillemots and puffins. These heavy-bodied birds are perfectly adapted to swimming underwater, and dive beneath or to the edge of the bait ball in order to corral it against the water surface.

Thieves are less subtle in their feeding strategy. Consisting primarily of a group of birds known as jaegers (and to a lesser extent gulls), thieves prefer not to catch their own prey but to steal it from others.

Suppressors can be the most destructive of all. This group includes birds such as shearwaters that plunge from the sky or swim through the centre of a bait ball to catch their prey. These disruptive feeding tactics scatter the fish, driving them out of range of the other flock members, and dispersing the bait ball.

Predators from above (above): gulls make a raucous descent on any herring that come to the surface, whether they are spawning in the shallows or trapped in fishing nets.

Spring

As the sea warms and wakes up in March and April it heralds the start of the new season in the Pacific North-West, which is marked by a truly remarkable natural event. The planet rolls on its axis, a strengthening sun causes sea temperatures to rise, and millions of herrings emerge from the deep water where they have spent the long winter. These vast shoals are heading to the shallows in order to spawn. Wildlife and fishermen eagerly await their arrival and for two years, so did our film crew.

As the shoals come into the shallows, gulls and seabirds gather to feed on both the herring and the roe. The roe's high calorific value is just the job after a winter of slim pickings. The birds snatch herring as they surface close to shore or get trapped in rock pools on the receding tide. Bald eagles swoop in to pluck fish from the water, or just stand on the water's edge, grabbing those that come within talon-reaching range. Alaska has the highest numbers of bald eagles in the world and during spawning it's not uncommon to see 30 birds along a small stretch of coast. Steller sea lions smash into the swathes of herring as they approach the shore, beating up the shoals and taking advantage of the sudden glut of food. And humpback whales that have either remained through the winter or already made the migration back from Hawaii are gifted with mouthfuls of fish, scooping them out of the shoals that blanket the coast. Underwater, fish, crabs, and other marine invertebrates gorge themselves on the bounty of eggs. It is a healthy first course on the menu of the great feast and a time for all to feed.

But despite the predators, the herring just keep coming. And for a few days at the end of March or the beginning of April the shallow waters are turned white with spawn. It's a phenomenal spectacle, its extent best appreciated from the air. Spawning is timed to allow the egg hatching to coincide with the plankton blooms. And although it is a feast for the predators, the sheer numbers of the herring and the trillions of eggs produced overwhelm the numbers of predators that are drawn in. These spawning events happen at various points along the coast, where different stretches of coastline are home to different herring stocks.

Doing justice to this spectacle meant capturing the action from every angle. While producer Hugh Pearson and cameraman Simon Werry covered the action from a helicopter, I worked at showing what was happening underwater. This turned out to be a rather smelly and sticky affair! Once the fish are in the shallows they appear to pulse along the shoreline in waves, the males first releasing their milt into the water, turning the water into a soup of sperm, then the females depositing their eggs on any suitably sturdy object, be it kelp, rocky seabed, or you!

An avian icon (left): Alaska boasts the world's largest resident population of the bald eagle, for whom the spawning event is one of the first good feeding opportunities after the bleak winter. Adults show the piebald colouring (from where the name bald eagle originates) of white head and tail feathers with a black body, while juveniles are a mottled brown.

In amongst the herrings

One technique we employed in order to film without disturbing the fish was to position a remote underwater camera on a tripod ahead of the spawning front. I swam down and sited the camera, feeding a monitor cable back to our boat. After a few minutes the fish arrived. For half an hour we'd watch on the monitor as wave after wave of fish pulsed through shot, each sweep turning the water whiter and whiter with milt, until the image turned to a whiteout. The shot had finished but the camera still needed to be retrieved. Diving into the spawn was like swimming in skimmed milk. Unable to see, I had to guide myself hand over hand back down the monitor cable, fumble around to find the camera, and swim with it back to the surface. All the while I could feel the female herring bumping into me, laying their eggs. There were eggs all over my drysuit, in my beard, all over the camera. The eggs were incredibly sticky, which meant that despite a good scrub in the lodge later, my dive gear smelled like a tropical fishmonger's two days after a power cut. Definitely not the glamorous side to filming wildlife!

Despite being rather smelly, the herring roe is highly prized and considered a delicacy. Since time immemorial, native people throughout the region have harvested it by placing spruce and hemlock branches in the shallows during the spawning to retrieve the eggs. A commercial herring roe fishery grew up in the 1960s to supply a lucrative Japanese market. The fish are seine netted prior to spawning, and transferred first to larger tenders, then onto freighters which transport the catch to Japan.

Safety in numbers (above): during the "event" millions of herring will crowd in to the shallows to spawn and overwhelm the numbers of predators that have gathered to feed.

Spawn to the right of you... (left): in the warmth of spring, hundreds of kilometres of coastline are turned white with herring spawn. There are three major herring stocks that are fished in Alaska, including the one pictured here in Sitka.

Herring fishery

The multi-million dollar herring fishery of south-east Alaska is one of the most lucrative and best-managed fisheries in the world.

Each year in spring, the Alaskan Fish and Game Department carry out daily aerial surveys to monitor the fish stocks moving into shallower water. This enables them to estimate the total numbers of fish present, and thus designate a small percentage of this stock as catchable quota for the year.

When the go-ahead to open the fishery is given, a sense of intense competition descends over the fleet. This fishery is split into just a few short openings in which the boats can fish (some lasting only 15 minutes each), and they're further restricted by being confined to a small area. As the countdown commences, the crews know they have only a short time to find a school of herring, cast their nets, and then close them before the time is up – and if their nets aren't closed before the countdown hits zero they will have to let their entire catch go.

In the excitement it's not unusual for boats to bump into each other or get caught in each other's nets. The reason for the high-octane nature of this event is the value of the fishery – 2008 saw the largest catch on record, with one boat netting 1.5 million tonnes in an afternoon. Yet despite such large numbers of herring caught, the fishery has been sustainably harvested in this way for years. In the height of the season, nearly 80km (50 miles) of coastline are turned white with the spawn of herring that escaped the nets.

A valuable crop (top): herring roe only 1–2mm($\frac{1}{25}$–$\frac{2}{25}$in) in diameter covers the kelp and rocky shore. It is prized as traditional food by local native peoples and by the Japanese.

Seine fishing (above): a seine boat "purses" its well-laden net to draw in the catch during the Sitka sack roe fishery.

The eggs are separated from the females and the carcasses ground to make fish meal. There is a lot of money to be made. We dived in and around the nets of a very generous fisherman, Jamie Ross, to film the herring being caught. To be amongst 40 tonnes of fish as the seine net is being drawn in is like diving in an oversized, fish-filled jacuzzi with the jets set to maximum. You are pummelled by a blanket of silver as the fish are drawn together. My exhaled bubbles burst through the body of fish, creating a window through which to look up at the nets as they closed tighter around the school. As the bubbles rose to the surface, the window closed as the fish crowded back and engulfed me. Being in the dark midst of all those bodies is the closest I've come to seeing life from a herring's point of view.

The new life of spring

As spring gives way to summer, the richness of life in the seas increases. With the lengthening days the greater amount of light produces the first phytoplankton bloom. This fuels a growing body of zooplankton that in turn feeds the shoals of herring, sandlance, and other larger fish. While we may not witness this growth of life directly, we can see the effect it has on the region's seabird populations. The variety of seabirds seen along this coast testifies to the richness of the system and the niches that are created, while the size of the bird colonies such as those on St Lazaria Island illustrate the abundance of fish food available. From May the small island comes alive with 500,000 breeding birds. Common murres (also known as

common guillemots), tufted puffins, and rhinoceros auklets burrow into the tussock-grass slopes, while gulls and cormorants take up residence on the cliffs. And the activity carries on around the clock. Fork-tailed and Leach's storm petrels descend under cover of night to nest amongst the tussock and Sitka spruce.

It was while filming in British Columbia that I had my most memorable encounter with these birds. We were woken by strange sounds in the middle of an inky black, moonless night as our boat rode anchor near Hope Island, off the northern tip of Vancouver Island. As we came out on deck, the night sky was electrified by the croaking screeches of Leach's storm petrels on their approach to land. In an otherwise silent sea we stood on the top deck, being buzzed by hundreds of mini spirits, which we could only glimpse as they whistled through the weak light emitted from our navigation lamps. The white anchoring light on the ship's mast had attracted these dainty birds, bringing them colliding with the boat, crash landing on deck. Storm petrels are truly birds of the sea and their small, clawed feet, perfect for picking food from the wave tops, are ill adapted to the slippery decks of a boat. They scrabbled around, struggling to take off, and we spent a good hour working along the deck, picking them up, holding them up into the air, and releasing them back into the dark. These birds spend the majority of their lives on the open ocean and only visit these remote breeding islands at night, the darkness protecting them from predation by gulls and skuas.

This time of the year is also the breeding season for other creatures reliant on the sea's bounty. Steller sea lions breed between mid-May and mid-July and they too choose the protection from predators offered by the more remote islands. They opt for rocky shorelines close to deeper water, giving them access to good fishing. We filmed on a number of haul-outs

"Flying underwater" (above): a group of common murres dive in pursuit of herring. They are well adapted to chasing fish, regularly feeding on shoals at depths of between 20 and 40m (65–130ft) but they have been recorded diving to 180m (600ft).

and rockeries, but the most windy and barren was Triangle Island, off the northern tip of Vancouver Island, a place that few people visit but that thousands of Steller sea lions call home. The storms that batter the island roll right off the Pacific, building in the open ocean to break themselves against the shore and on top of the sea lion colonies. A nine-month gestation and a three-month delayed implantation mean that courtship, breeding, and giving birth happen at a similar time of the year, which means the sea lions, and especially the dominant males, turn into landlubbers just when the sea is becoming productive. This might appear counterintuitive, but it ensures that the newborns have a full summer to fatten up ahead of their first winter.

Stellers underwater

The month that our camera team spent on the island coincided with the peak pupping time. The Steller's feisty nature starts from an early age. Pups, already battling in the hierarchy of the colony, squabble in the pools, while bulls contend for harems of females, defending patches of shoreline in titanic clashes. Nothing quite prepares you for the smell. To be alongside hundreds of animals farting, pooing, and peeing as they lie about, digesting their very rich fish diet, is a memorable sensory experience, to say the least. And as on land these are shy animals, in order not to disturb them it is best to approach them from down wind!

Even in these milder months large storms can sweep the seas, and the aftermath of the storm fronts and the dead bodies of newly born pups washed up on the beach really brought home the hardship of living in these seas. It is these storms and the nutrient-churning they create that make these seas for some so treacherous but for many so productive.

Approaching and filming Steller sea lions on land takes great care. As they are less at home and less confident on land we worked hard to avoid frightening them. Other considerations apart, we were after film of natural behaviour – images of scared or harassed animals are worthless. But Stellers are almost schizophrenic in the difference in their character on land and in the sea. In the water the game swings fully around and they are entirely in control. Here the sea lions are transformed into sleek, rubbery beings. They come spinning out of the green gloom, floppy torpedoes on a crash course that approach to within almost touching distance, then twist away at the last moment. The nervousness seen on land is replaced by boundless curiosity.

But this newfound confidence makes them surprisingly difficult to film doing anything "natural". You've had it the moment they've noticed you. Over they come, marauding in to eyeball the camera, dancing shapes in the reflective domed front of the lens. They chew the camera cables or mouth your fins, arms, or head. And the numbers build until you're looking at a wall of whiskered faces, big-eyed stares fixated on you. Then at times they relax and seem less distracted by your presence. They may start socializing, greeting one another snout to snout or touching a larger or more dominant animal. They may also embark on their curious behaviour of swallowing pebbles that stay in the stomach and are thought to aid them with buoyancy control.

With time we managed to get snippets of natural moments as we sat on the seabed, witnesses to a world that few get to see: the balletic grace of a 1 tonne bull, transformed from a lumbering giant to a heavyweight corkscrew, its powerful shoulders and 1m (3ft) long flippers slickly pulling it through the water. Or the gentle, reassuring touches amongst a mob of juveniles, bickering like teenagers on land but becoming graceful underwater and together, in an instant, with a few fin strokes, disappearing into the gloom.

Furry defences (above): sea otters are the only marine mammals along this coast not to be protected from the cold waters by an insulating layer of blubber. They survive instead by having the thickest fur in the animal kingdom.

At the top of the table

The presence of so many types of marine mammals in these waters is yet another illustration of the complexity of the ecosystem. As well as Steller sea lions, you can expect to encounter seals, sea otters, Dahl's and common porpoises, humpback and minke whales. You might even spot a northern elephant seal or a grey whale. Up until about 120 years ago you may also have been fortunate enough to see a Steller sea cow, the largest species of the order Sirenia, which has since sadly been hunted to

extinction. But for me the most exhilarating marine mammal to encounter is the orca or killer whale.

The orcas' intriguing beauty belies their intelligence. The black and white counter-shading pattern, like an art deco design, looks almost painted on in its vividness. Sharp fin shapes and smooth bodies give the whales a real presence when they break the surface – they are one of those animals that are just incredibly exciting to see. But hidden beneath their perfectly sculpted lines and patterns lies an intellect that can't be seen in facial expressions or gestures, but can be understood through witnessing their behaviour.

Killer whales show a variety of ways of coping with their changing food supplies. Some resident pods stay put and live only on fish, some roam the coast looking for marine mammals, while others just call in from the open

Killer whale
Orcinus orca

The largest species of the dolphin family, killer whales are now known to be separated into three distinct types that are considered to be sub-species or possibly even species.

Residents feed almost solely on fish and live in complex family groups, often with a dominant female at the head travelling with all of her dependent and adult offspring. Both male and female offspring can stay with the group for their entire lives, only leaving the pod for short periods of time to mate. Residents vocalize in sophisticated and highly variable "pod-specific dialects".

Transients by contrast feed almost exclusively on marine mammals. They generally travel in smaller groups (2–6 individuals) that roam widely up and down the coast, and don't always stay together as a family unit, with individuals regularly coming and going. They are much less vocal than residents, which is thought to be because their mammalian prey has good hearing underwater, and these orca prefer to hunt in stealth mode.

Little is known about the last group, known as oceanics, other than that they cruise the open ocean and are thought to feed on fish, sharks, and sea turtles.

KEY FACTS

Size	up to 9.6m (30ft).
Weight	up to 8 tonnes.
Intelligence	orca exhibit complex social behaviour and learned hunting techniques that are passed down from generation to generation in what some scientists describe as culture.

Resident orcas (left): killer whales of this type are very communicative underwater, as their staple prey of salmon is unable to hear their calls – unlike the marine mammals hunted by transient orca. Each resident family group has its own signature calls.

ocean. After humans it is killer whales that as a single species have spread to more areas and adapted to more situations than any other creature on our planet. The variety of their feeding strategies is one illustration of how intelligent and adaptable they are. And it is in this part of Canada that so much has been observed and discovered about their way of life. The most recent observations by two eminent whale biologists, John Ford and Graham Ellis, reveal that different resident pods become specialist feeders concentrating on salmon of different species and ages. At the times of the year when salmon are most abundant, the orcas deliberately hunt only the individual fish with the highest fat content, cherry-picking the most nutritious meal.

Such insights come only after years of dedicated and patient research, but encounters during our filming hinted at the killer whale's intelligence. One glassy calm day in south-east Alaska stands out. We had been filming underwater near a sea lion haul-out for a number of days. At first Shane and dive guide Bryan Gundaker had been characteristically mobbed by the sea lions, but on the third day the animals were incredibly skittish, even underwater. They hugged the coast, lingering shyly in the shallows. In order not to disturb them, Shane mounted the camera on an underwater tripod and backed off into deeper water. In a whirl the sea lions disappeared, shooting to the rocks in panic.

As Shane recounts, the penny suddenly dropped – it wasn't him that sea lions were wary of. As he slowly turned around he saw three transient orcas looking at him. As Shane edged towards his camera the whales moved back into the gloom. No shots, but one of those encounters that lives with you. Once back on board we shadowed the orcas along the coast at a distance to see what the encounter might bring. But rather that us going to them, two younger animals became curious about us. As we gently motored through the calm water, the whales bee-lined towards the ship's stern, coming within a few metres. They hung there in the water, looking at the propeller, apparently intrigued by this thing that was making strange noises in their underwater world. Once they left we were all exuberant to have been buzzed by orca. And amazingly, after 10 minutes they were back. They did this a total of four times, giving us something our salty ol' local captain, Scott McLoed, understatedly described as a "pretty good day".

Orcas in action

Spending time on location you witness the beauty of the natural world, but you also see things that can shock and disturb. We came across one such scene while en route to film at a Steller sea lion haul-out. Angela Smith, our wildlife guide and skipper, spotted tall black fins on the horizon and instinctively knew that something wasn't quite right. As we headed

A family group (above): the orca's social structure is matrilineal, with a group comprising two or more related females and their offspring. In resident orcas, both male and female offspring stay with their mothers for their entire lives.

offshore to investigate we stood in the open inflatable, trying to make sense of the scene in front of us. Tangled amongst a raft of kelp lay an animal, but what was it – a sea otter, a seal? As we approached we saw it to be a bull sea lion, lying on its back, but it was far from well. Its belly was bloated and it hung uncomfortably in the water, labouring to move and struggling to breathe. Four black fins broke the surface nearby – a pod of transient killer whales, marine mammal-hunters.

The sea lion twisted in the water trying to keep watch on his attackers. It looked as if his back was broken and he had lost the use of his hind flippers. Nevertheless his jaws still posed a real threat to the orcas. As we watched, in they came, one or two at a time, sliding up next to the sea lion to smash him with tail blows. Even now I can hear the sound of those deadening thuds and the booming splash of the water exploding. The sea lion twisted to stay on his back, perhaps using the surface of the water to protect his tender belly, perhaps to see more easily from where the orcas would approach next. The killer whales made pass after pass, knocking and slapping the bull, injuring him internally. At times they'd come in as a pair, one sliding through as if to distract him while another swam in from a different angle, ramming him with a nose or another tail blow. One of the orcas was a youngster, and would shadow another as it swam in, as if trying to learn their methodical technique.

The orcas kept up their assault for over an hour, slowly wearing the bull down. Eventually they pulled him underwater. For a time they surfaced two together, leaving two to share the feeding out of sight in the depths. I've since learned that it is thought they feed as a pair, with one holding the prey while the other takes bites of flesh. Gulls, appearing as if out of thin air, scavenged morsels that drifted up to the sea surface. Once all four killer whales surfaced together we knew the feeding had finished and the sea lion had gone. The orcas travelled on. The birds left, and only the quiet sea bore witness to the most powerful thing I've yet to witness in the natural world.

Co-operative killers (right): cameramen Shane Moore and David Reichert film the transient orca group as they hunt a male Steller sea lion.

Deep-sea colour: in an otherwise emerald green sea, populated largely by dull-coloured animals, the fish-eating sea anemone *Urticina piscivora* **(above left)** and the purple-ring top snail *Calliostoma annulatum* **(above right)** provide an unexpected riot of colour as they nestle amid the kelp.

It is impossible to watch something like that and remain unaffected. How did they finally drown and kill that bull? How could they break up such a tough creature's body? Maybe there are some things best left to the imagination, things that we are never likely to see, in the depths of the ocean.

Summer

As the year continues and we enter the summer months, the seas become increasingly alive. But ironically this caused us real difficulties with filming, in the form of severely reduced underwater visibility. At the height of the plankton bloom I found myself floating in an emerald green soup. Filming larger creatures such as sea lions in that green gloom proved very hard and frustrating – though for the shots we needed where plankton was the subject it was perfect! The previously withered and storm-washed bull kelp had flourished into a forest, with thick fronds having grown at a rate of up to 30cm (10in) a day. It was September and we were filming the underwater walls of Browning Pass in British Columbia, where carpets of anemones and barnacles sift and net the plankton that is swept in on the current. On such busy walls of life, concentrating on the smaller animals can be as enthralling as looking for the larger creatures. As I watched, a barnacle pushed out a long cirral tentacle, twisting it across the current to sweep the water for morsels of food. Having once, in its larval life, been part of the plankton, the barnacle was now firmly rooted on the wall.

In North America everything seems big and the invertebrate life is no exception. These enormous barnacles grow to 8cm (over 3in) in length in the wave-washed intertidal zone, while the carpets of brilliantly coloured anemones spread deeper down the rock walls. In these same walls you can find the giant Pacific octopus which, with a span of up to 6m (20ft) from tentacle tip to tentacle tip, can grow to over 45kg (100lb) in its three-to-five year life span.

The strength of the current ensures that a regular supply of nutrients is brought in to the creatures that fix themselves to these underwater walls. While the large variation in plankton levels happens over the course of the year, changes in currents are related to the tides and so can be seen on a daily and monthly basis. Filming underwater in currents that can whip up within minutes means working within the windows in the tides. As is usually the case wherever we film, local knowledge combined with a good tide table is the best reference. Whirlpools and eddies can suddenly appear, churning up the sea surface. On one dive in British Columbia, cameraman David Reichert got sucked into a whirlpool while he was engrossed in filming. He had started at 3m (10ft), but at 30m (100ft) decided it was time to stop and get back to the surface. These whirlpools just as quickly, almost instantly, abate when the currents calm down as the moon drifts on it path and the tide it creates reduces.

A sea of fish

At the height of the summer these waters truly feel busy with life. A crucial element in the ecosystem is the fish that supports the wealth of larger creatures, and one fish in particular is top of the menu – herring. The herrings and fish like sandlance are the link between the rich plankton resources and the larger creatures in the marine food chain.

Underwater feeding (right): tentacles emerge from a group of gooseneck barnacles (*Pollicipes cornocpia*). Unlike other barnacles, which use only their tentacles to feed in the current, a gooseneck barnacle's long stalk means its whole body sways in the current and its delicate tentacles don't have stretch so far to fetch food.

Bubble netting (above): a group of humpbacks, mouths agape as they rise out of the water. It is thought that these whales are not related to each other but come together solely for the convenience of co-operative feeding.

The zooplankton (the collection of small creatures that live in the water column) feed the herring shoals, and at this time of year it is these shoals that attract a wealth of predators.

A visual illustration of the importance of the herring can be seen in certain places along the coast during the months of August and September. A concentration of fish near the surface creates a bait ball that draws in seabirds, seals, and whales to feed in a frenzy of activity. This concentration of life was something that we wanted to film. Bait ball activity appears to be greater during the higher tides, perhaps due to the increased turbulence created as more water passes over the underwater peaks and valleys, bringing up pockets of water and with it shoals of herring. As the fish are drawn up nearer the surface they come into the range of the diving seabirds, in particular the common murres, the architects of the bait balls. It is these alcids – true seabirds that only ever come ashore to breed – that first find the herrings and start working them up to the surface. As David and I lay on the surface, hundreds of murres and auklets streamed past beneath us, effortlessly crisscrossing under the herrings as if flying underwater. As the number of birds increases, the herring shoal concentrates into a ball for protection. The murres coax the ball to the surface, each bird taking care to swim under it and only pick off fish from beneath or grab the odd wayward straggler.

This tightly packed, shimmering ball of fish, the size of a truck, is gradually reduced as rafts of excited birds gather on the surface and feed. Cormorants, loons, and auklets all make the most of the murres' hard work. But it's the gulls that really ransack the bait ball in a raucous, free-for-all, party-crashing approach. The herrings come under full assault, jabbing beaks raining down from above as an army of murres pick at them from below. A break in the birds' onslaught sometimes seems to give the herrings an escape route and they stream away from the birds. But again the murres shoot in front to head them off and re-form the ball to continue feeding. As the birds feed and the size of the bait ball dwindles, the water fills with silver scales that drift down like snow.

We weren't the only ones to have worked out that it was the birds creating the bait balls. The local humpback whales appeared to have learned this too: on a number of occasions we saw a humpback swim in and snatch the entire bait ball from the hard-working murres. It wasn't much of a contest between a 1kg (2lb) bird and a 40 tonne Goliath. When the whale burst through to engulf the bait ball it indeed produced an amazing shot – and I believe a world first – but for cameramen David and Shane it was almost too close for comfort!

The summer sensation

It is the presence of humpback whales at the height of the summer that typifies how productive the waters of the Pacific North-West become. The humpbacks' way of coping with the lean winter months is straightforward – they simply leave. Amazingly they swim all the way to Hawaii, a round trip of over 14,500km (9,000 miles), but return here to feed. In this "feast or famine" lifestyle, summer is the key time for gorging. The humpbacks' round-the-clock feeding offers one of the most extraordinary spectacles in the whale world, and it was what we'd come to south-east Alaska to film.

To stand any chance of filming the whales we spent long days on the water, covering a lot of distance in search of the pods. For three weeks we cruised the waterways of Chatham Strait and Frederic Sound, running our tug for 300 hours, constantly on the look out. In the height of the summer Alaska nights are short, so filming days start at 4.30am. We tried to keep a rolling watch, one team member on the helm, one spotting, leaving the third to rest or more often than not brew the coffee.

As you scan the horizon, the blow is typically what you first see. In the right light you can spot the plumes of breath from miles out and quickly get an idea of the size of the group. On still mornings the sound of the blow can carry far across the water. It's usually the characteristic noise of a strong exhalation, but occasionally individuals stand out, some with rasping breaths, others sounding like foghorns. As usually only the whale's hump and tail are seen from the boat, it is the sound of the breath, the air resonating as it travels through huge tubes, that gives a clue to the massive size of the lungs, and the sheer scale of the creature beneath the surface.

Humpback whale breaching (right): no one really knows why whales do this, but it is thought to be the result of strong emotion – whether anger, happiness or the excitement of courtship. Whatever the reason, it requires a lot of energy to lift its 40 tonne body clear of the water.

That size is revealed when the whales breach. They seem to do this when stimulated, for instance after a feeding session, but whether it is out of excitement, contentment, frustration, or joy, we don't really know. Watching a breach is like seeing a living torpedo erupt from the surface, albeit a rather fat one. As the snout pierces the surface, a wave of water curls back along the body, and when fully emerged the whale half twists in mid-air, falling with a tremendous flop back into the sea, which explodes with a double splash. Sometimes an individual will breach as many as 30 times without pause. It's a monumental effort to heave their great bulk from the water and it must use a lot of valuable energy, illustrating how well the whales are feeding at this time of the year.

Humpback whales use a number of feeding strategies to harvest these rich waters, but by far the most impressive is the cooperative feeding

Humpback whale
Megaptera novaeangliae

The humpback is one of the rorqual whales, a family that also includes the blue and minke whales. It can be easily distinguished from other Cetaceans by its extremely long pectoral fins, which can reach up to 30 per cent of the animal's length. The shape and colour pattern of these dorsal fins and of the humpback's tail flukes are unique to each individual.

Humpbacks are baleen whales, meaning they have a series of 270–400 fringed overlapping plates hanging from either side of the upper jaw, made from a hair-like material called keratin. During feeding, characteristic ventral pleats on the whales' throats expand, allowing them to take on huge volumes of water which is then expelled through the baleen plates, filtering food from the seawater.

KEY FACTS

Size	females larger than males, up to 15m (50ft) long.
Weight	up to 40 tonnes
Young	one calf every 2–3 years. The young average over 1 tonne at birth and measure about 3.5m (12ft).
Food	various small fish and tiny shrimp-like crustaceans called krill; during their peak feeding time they can consume an astonishing 1.5 tonnes of food a day.

South for the winter (right): a humpback whale in the clear but unproductive winter waters of Hawaii, at the southernmost point of its extraordinary migration.

Whale migration and mating

Although a few humpbacks are seen in south-east Alaska throughout the year, most undertake an epic migration at the end of summer. In one of the longest migrations of any mammal, they leave their feeding grounds in Alaska and swim 5000km (3000 miles) across the open ocean to the warm tropical waters of Hawaii, where they mate and give birth. Although the length of time needed to complete this migration is not well documented, one whale was seen in Hawaii a mere 39 days after it was spotted in Alaska.

Although the waters of Hawaii are warm and sheltered, they contain little food, and so once the calves are strong enough the whales migrate back to Alaska, arriving around April in order to cash in on the abundance of food brought about by the plankton bloom and the subsequent herring spawning. How they navigate back to the same spots each year remains a mystery.

Deep blue sea (above): a humpback mother and calf, Hawaii.

behaviour for herrings, known as bubble netting. It is remarkable in terms both of the sheer spectacle and of its indication of the intelligence and skill that enables the whales to fish for herrings using bubbles, sound, and the sea's surface. This behaviour was first described in scientific papers by Cynthia D'Vincent and Scott Baker in the 1980s. More recent work by Fred Sharpe and others continues to shed new light on this amazing behaviour, including showing that groups of feeding humpbacks aren't related and come together only to feed cooperatively like this at certain times of year.

Filming whales is all about patience, prediction, and good boat etiquette. Predicting where they'll come up is partly luck, partly judgement. The direction that the pod is moving can be worked out by the following the direction of the last fluke down, but guessing how long they'll stay down is where the luck comes in. As you're tracking at a distance, it's vital to keep the boat's speed low and constant. Revving engines or constantly changing speeds creates sonic mayhem underwater, so a calm, careful approach is the only method. A special stabilized camera mount fitted onto

Summer plenty (above): for humpbacks, as for most other Alaskan animals, summer is the most productive time for feeding. The whales travel like this for many kilometres, constantly on the look-out for shoals of herring or swarms of krill.

the boat allowed us to get steady shots without encroaching on the whales. When busily feeding, the whales stick to the same depth contour and work along the coast. Often we found them at rocky outcrops close to shore or in bays where perhaps the herrings get sucked into eddies that develop along the coast or are swept up from the depths. The steepness of the shoreline underwater in these fjord channels is hard to appreciate unless you can see a depth sounder, or examine a chart. The abrupt flanks of the surrounding mountains carry on underwater, and within a stone's throw of the shore, the sea bed can fall to over 60m (200ft), tumbling away to great depths in the centre of Chatham Strait. The presence of these underwater mountains mean the whales come in incredibly close to shore, at times working no more than 10m (30ft) from the rocks. When the pod consistently feeds along a contour it gives the best opportunity to get good footage of the bubble netting.

The whales dive as a team. One by one, backs arch and they slip beneath the surface. Tails out of the water indicate that they've dived for a while, so we let the boat gently move forward and then cut the engine to listen. It's the suspense of waiting for the whales that makes watching bubble netting so exciting. From beneath the water comes the muffled moan of a single humpback – the net is being laid. The underwater sound is so loud you can hear it when you are standing on the boat. The moan gets louder and louder, and all eyes are now scanning the sea, looking for the first bubbles to break the surface. "Two o'clock!" The net is emerging to the right of the bow and Shane swings the camera around to frame up on the centre of the ring of bubbles. Camera running. The water simmers as chased herrings appear to dance on the surface. Boom! Eight humpback whales explode from the sea, open mouths lifted from the water. As they sink back down, mouths close and throats expand. The pink pleats under their jaws are stretched open to allow as big a mouthful of water and fish as possible. As the whales loll on the surface, pushing out the water through baleen-lined jaws to leave a mouthful of fish, gulls crowd in to grab any herrings that have escaped. The whales then start moving once more, back into formation, and soon dive again, while Shane and I are left on deck wondering what it must feel like to have a belly full of wriggling live fish.

Humpbacks are known as the singing whales and have a rich vocabulary of sounds. The calls used for bubble netting are not only unique to this behaviour, but also breathtakingly beautiful. I'd taken along a fantastic high-quality DPA hydrophone – an underwater microphone. Nothing quite prepares you for the first time you put the headphones on. A siren-like call fills your ears, a rich, droning note like an aquatic cello.

Aerial view of bubble netting (right): the whales reach the surface in the centre of the net. Such groups may consist of between two and fifteen whales, and it is thought they organize themselves in the same order on each attempt, perhaps based on a feeding hierarchy.

One whale is singing, at times accompanied by a second, apparently harmonizing its notes. In the depths the whales are starting to move, running the ring of bubbles beneath the shoal of herrings. The pitch creeps up and with it the sense of suspense. Higher and higher rides the pitch, as the team of whales rises in the water. Suddenly the pitch jumps into a scream – the stun call – and moments later the whales break the surface. The stun call is thought to confuse the herrings in the final throes of the netting, and the song is believed to chase the herring into the net and help synchronize the team of whales to fish the net together. But why we should find the call so beautiful is a mystery. It's a child-like sensation to experience something so fresh and so unrelated to anything else you know. I can hear the song in my head as I write this and at the time the crew and I sat for hours sonically spying in on the humpbacks' world.

The end of the feast

September drifting into October marks the end of the summer in these northern waters. The daily amount of light diminishes, the autumn storms become more frequent, and the productivity of the sea drops dramatically.

Visiting summer birds fly south while the humpback whales start their epic swim to the warm waters of Hawaii. The feast has finished. The plankton bloomed, their nutrients consumed and passed up through the food chain, and the wildlife that remains faces the long, lean winter months before the system rejuvenates once again in the spring. For all their summer rewards these waters are a difficult and challenging place to live, and for filmmaking it's an equally tough place to work.

But perhaps that's only to be expected in such a true wilderness. In so many places that we now travel to and film in, wild areas are increasingly being hemmed in by constantly encroaching "development"; in the 21st century human-tainted landscapes are becoming the norm, leaving only occasional pockets that can truly be described as wild areas. But here in Alaska and in the remoter parts of Canada the opposite still applies; the constant is wilderness and nature, and human influence is limited to a handful of urban areas. This really is the "last frontier", as the Alaskan state licence plates testify. It is a place that gets under your skin, and, for me, is home to some of the best and most rewarding wildlife experiences in the world.

One of the last wildernesses (above): the mountains and islands of south-east Alaska's Tongass National Forest continue to hold the biggest tracts of virgin old-growth forest left anywhere in the United States. The seas that surround them remain a haven for wildlife.

The Great Migration The Serengeti

Imagine this scene. Thousands of antelope are tightly packed on the shores of a lake. They're on the move and the smell of dust that rises from their hooves pounding the dry ground is almost overwhelming. The dense crowds unfold as far as the eye can see – and beyond. Isolate a group and the horns and bodies of the animals take on the strange abstract beauty of an Escher print. These are the white-bearded wildebeest of eastern Africa, and en masse they create one of the great events in nature. But such a spectacle is only part of a fascinating story.

Migration is the way of life for *Connochaetes taurinus*. They are the most numerous antelope in east Africa and travel through some of the most exciting landscapes in the world. From the lofty heights of the Masai Mara plains in Kenya down to the lower altitudes of the Serengeti in Tanzania, they cover an area of 65,000–78,000sq km (25,000–30,000 square miles) over the course of a year in a constant search for fresh grazing. Their journey takes them in a haphazardly circular pattern from south to west to north, then back south, out to the south-east, and back again. These annual movements define the Serengeti ecosystem.

The Serengeti
East Africa

Together, the Tanzanians and Kenyans have put aside an area roughly the size of southern England to safeguard the traditional migration routes of the white-bearded wildebeest.

This means that all associated wildlife such as lions, hyaenas, cheetahs, leopards, gazelles, buffalo, giraffe, elephants, and a host of smaller animals have been provided with a safe refuge. It also ensures that some of the most spectacular landscapes in the world are protected, including areas rich in hominid fossils. Since these countries are among the poorest in the world, their commitment to set land aside for conservation sets a standard that most western countries would find hard to emulate.

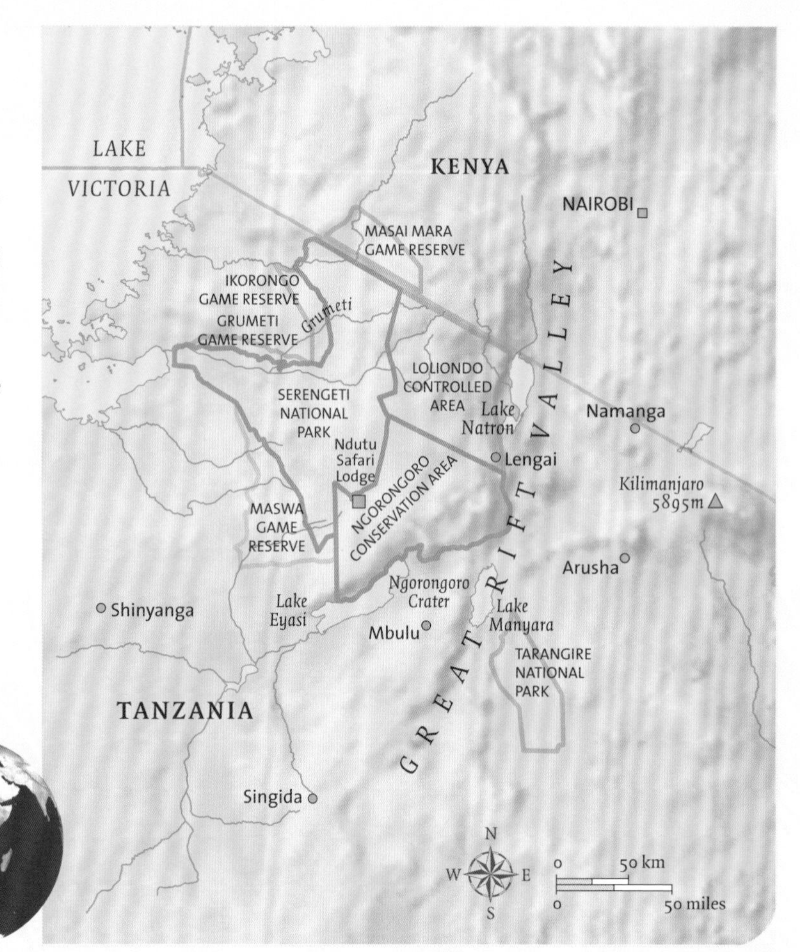

Fresh feeding grounds (right): as the wildebeest move across the plains, the pattern of the herd from the air briefly takes on the form of a breaking wave on the shore. In a few hours, though, they disperse and the moment is lost.

Wandering the plains

The antelopes have no respect for manmade boundaries that exist purely on paper. In Tanzania, they roam through the Serengeti National Park, Ngorongoro Conservation Area, four game reserves (Ikorongo, Grumeti, Maswa, and Loliondo), and one game-controlled area (Natron). Across the international border in Kenya, they travel in and out of the Masai Mara Game Reserve. In all these places, the landscape is part of, or is dominated by, the Great Rift. A complex system of faults, escarpment, and volcanoes

Ol Doinyo Lengai (left): the active volcano seen in all its glory with the knife-edged ridges emanating from Empakai crater in the foreground. The volcanic highlands form a dramatic backdrop to the southern Serengeti Plains.

forming the walls of a wide valley, this geological feature is clearly visible from space and cuts a swathe through Africa for 6500km (4000 miles), all the way from Lebanon to the Mozambique Channel. Imperceptibly to our eyes, the rift is widening and will, eventually, tear Africa apart. But in the meantime, the geological movements have created a variety of habitats, including the plains through which the wildebeest roam.

Both day and night in East Africa are pleasantly warm, but during June and July, night temperatures drop to around 10°C (50°F). The driest months are June through to October, by which time the land is parched and baking in a daytime temperature of 38°C (100°F). Rainfall is rarely widespread, with averages fluctuating greatly from one place to another. In Tanzania, the first few drops in November mark the beginning of the rains, though heavy showers can be few and far between. Historically, January, February, and March tend to be the wettest months, though in recent years the vagaries brought about by our changing climate have played havoc with rainfall patterns.

After seven months of eruption (above): the ash from Ol Doinyo Lengai carpets the floor and escarpment wall of the Great Rift Valley. Carried by the prevailing south-west wind, the ash was also wafted far out onto the Salei Plains.

Such unpredictability means the movements of the wildebeest are equally erratic. Spiralling costs dictated that we couldn't afford to film in Tanzania for longer than seven or eight months, so trying to decide which periods we should pick in order to get the footage we needed to tell the story was nerve-wracking . However, the choice of where to go when is even more vital for wildebeest. Driven by the need to find protein, vitamins, and water, each individual travels about 10km (6 miles) a day. The wrong decision is life threatening.

On average, the vast herds spend eight to nine months of the year in Tanzania and the rest up north in the Masai Mara. The rainfall there is twice as heavy as in the Serengeti Plains so, as the dry season bites in Tanzania during August to October, the wildebeest seek refuge in Kenya.

In November when the rain starts to fall in Tanzania, the wildebeest move south. By February, if the rains are good, they spread out over the

Wildebeest
Connochaetes taurinus

White-bearded wildebeest have broad muzzles adapted to close, rapid, bulk feeding on grass. Few tropical mammals have such a restricted birth season, with some 80–90 per cent of calves being born in the space of three weeks.

The calves are the most precocious in the world and have been seen standing and attempting to suckle just under three minutes after birth, though six minutes is the average. Each mother recognizes her calf by scent and rejects all others, and mother and baby also learn to identify each other's calls. It's estimated that when mothers and calves are separated – which happens a lot in the tumult of the giant herds – reunion is successful in about 90 per cent of cases.

KEY FACTS

Height up to 138cm (54in) at the shoulder.
Weight up to 274kg (600lb).
Density up to 270 per sq km (700 per square mile).
Birth about 90 per cent of females give birth within a few weeks, providing the predators with a glut of easy prey, but ensuring that a viable percentage of the calves survives.

Watery crossings (right): every year, the wildebeest herds cross and re-cross many small streams, gorges, and gullies, but one of the biggest obstacles they face is the Mara River.

No time to waste (above): a wildebeest calf struggles to its feet just a minute or two after its birth. You can't help but be impressed by its tenacity and strength.

Preparing for the rut (above): two male wildebeest lock horns in a tussle to test each other's strength in preparation for the battle to come.

A convenient perch (above): a wildebeest also provides wattled starlings with hair for their nests and food from insects stirred up by the herd's hooves.

Yearly migration

If wildebeest could stay put, they would. Instead the need to find high-quality nutrition takes them on a long annual journey from north to south and back again.

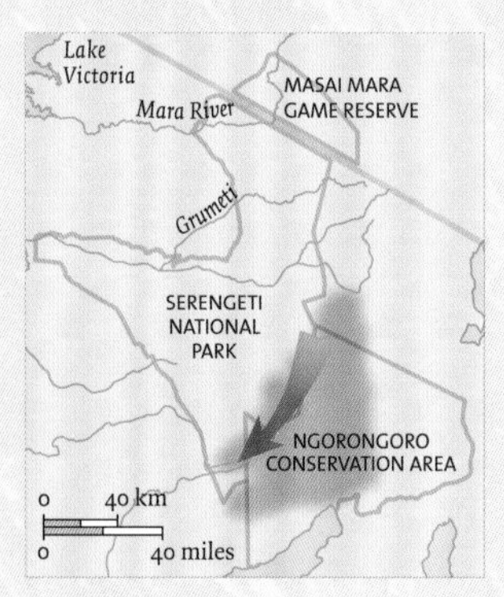

December to February

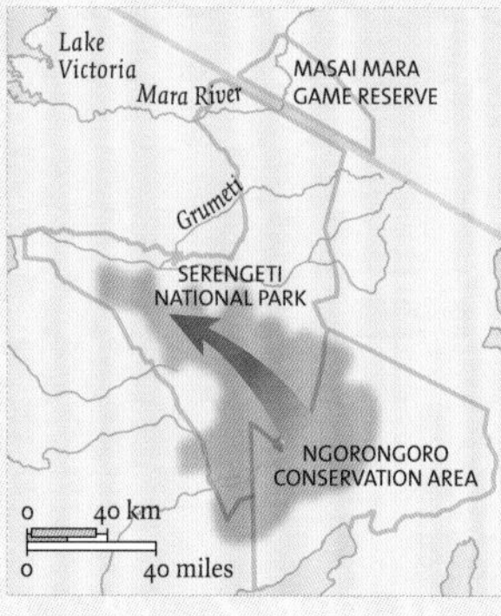

March to May

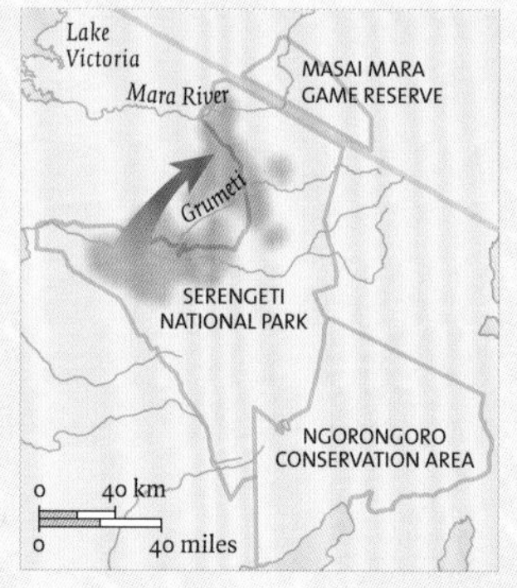

June to August

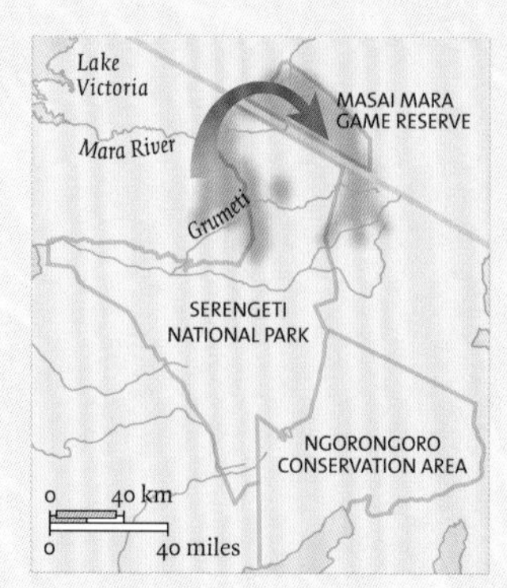

September to November

Serengeti Plains. Here a combination of mineral-rich volcanic ash and low rainfall has created grasses richer in phosphorus – a mineral important to fertility, appetite, milk yield, and growth – than anywhere else in their range. So if the rainfall is high enough, this is where the wildebeest have their calves. During February and March, 90 per cent of the female wildebeest give birth in just a few short weeks and the short-grass plains are inundated with thousands of newborn calves every day – truly one of the world's most amazing wildlife spectacles.

When the rain on the arid plains is limited, the wildebeest are forced to give birth in the thick whistling thorn thickets of the southern Serengeti. Trying to manoeuvre a car through this scrub and bush is so tricky that filming the wildebeest birth is almost impossible. However, the short-grass plains are an ideal location for both wildebeest and filmmakers. The local nomadic pastoralists, the Masai, call them *siringet*, which, translated into the more prosaic English phrase of "extended area", does little justice to the stark beauty of the grasslands. Wide, open, and windswept, they stretch to distant horizons – and beyond – with barely a tree in sight. But why is this?

The reason lies in the impressive line of volcanoes that rise up to the south-east of the short-grass plains. All these volcanoes are dormant or extinct bar one – the breathtaking Ol Doinyo Lengai, born as its neighbour, Keramasi, died about two million years ago. In its death throes, Keramasi's eruptions produced great clouds of lime-rich ash that the strong prevailing winds blew westward, where it blanketed the surrounding plains and mountains with deep drifts.

A river of life (above): wildebeest are inseparable from the story of the Serengeti, providing spectacle, beauty, and power wherever they go.

Characters of the Serengeti
The dry season

Caracal *Felis caracal*
Rarely filmed or even seen in the wild, caracals prefer to hide away during the day in thick, woody vegetation but are active at dusk and night, hunting for birds and rodents on open grassland.

Some years ago, we were lucky enough to spend three weeks watching a caracal with three kittens for a BBC documentary called *Cats under Serengeti Stars*. During this time, we filmed her leaping over 2m (6ft 6in) to catch a European white stork as well as sprinting after, and catching, a spring hare. While she was hunting, her kittens tucked themselves away in disused aardvark holes and one night we witnessed at first hand just how tough these little cats can be, when their mother had to stand her ground and drive five spotted hyaenas – all of which were twice her height and weight – away from the kittens' hiding place.

Aardvark *Orycteropus afer*
As the only surviving members of the *Tubulidentata* species, aardvarks are genuinely unique. To see one, especially during the day, is such a rare event that much of their lives and habits remain a mystery.

They need sandy or clay soils in which to dig for beetle larvae, termites, and ants. Finding these by smell, they trundle around with their nose close to the ground before making shallow, exploratory scrapes. If the food supply is plentiful, they dig deeper holes at rapid speed. They also excavate permanent shelters in which to rest during the day, some of which can be extensive. When unoccupied, burrows provide shelter and refuge for many other mammals, birds, reptiles, and insects.

Warthog *Phacochoerus aethiopicus*
Warthogs are able to survive the driest of dry seasons. This means they are the only pig adapted to life out on the plains and woodland edges of the Serengeti.

During the day, females and their young are active but seek safety underground well before dark, whereas males continue to forage for an hour or two longer. Highly conspicuous and present all year, warthogs become the favourite prey of lions in the dry season. When chased, they often seek refuge in holes, reversing into them so that they can use their tusks in defence. But lions are often so hungry that they spend hours trying to dig their prey out and their superior numbers mean that others can surround the pig and attack from the back.

Grass-eaters

Elephants and buffaloes pull and trample long grasses, stimulating the re-growth of finer grasses suitable for zebras and wildebeest. With top and bottom incisors, zebra can cope with tough grasses, whereas wildebeest, with incisors only on the bottom of the jaw and a hard biting pad on the top, prefer more tender grasses. Gazelles have narrow muzzles and favour the shortest grasses of all.

Masai and their cows

Although the Masai people now appear as the archetypal pastoralists of the Serengeti scene, they are relatively recent arrivals, driving out the Datoga people 150–200 years ago. However, hominid history of the area stretches as far back as 3.5 million years when our very remote ancient ancestors lived and hunted in a landscape very similar to that which we see today. In fact, some scientists believe the widespread and frequent use of fire by hominids actually created much of the savannah landscape, making it the oldest man-made habitat in the world.

Too heavy to be carried far by the wind, the soils nearest to Keramasi are coarse, sandy, and highly porous, while further away they are lighter and more finely textured. Over centuries, the rain has trickled through the ash, forming a calcrete layer just below the surface. This layer is so hard that no roots have the strength to penetrate. Since deep-rooted trees are prevented from getting a foothold here, shallow-rooted grasses are able to take centre stage.

Further away from Keramasi, the soil holds more moisture and is deeper and less salty, so different types of grasses can thrive. Here dwarf forms of dropseed and star grass create rolling plains as smooth as velvet. Further west again are rich stands of red-oat and bamboo grass that, with good rains, can grow as high as the backs of grazing animals. And so geology and ecology are inseparably linked in a tight cycle of fertile volcanic ash, rain, and grasses that results in a unique ecosystem.

Features of the landscape

Breaking up this grassy scene are beautiful rounded rock outcrops called kopjes (pronounced "copy", from the Dutch meaning "little head"). Made of ancient granite, they glow softly pink in the setting sun. Some are no more than single rocks; others are a collection of massive boulders providing shelter for an array of plants, trees, and wildlife.

Further west and north, breaking free from the grip on the landscape exerted by the volcanoes, trees dominate. Here are huge tracts of flat-topped acacia trees (*Acacia tortilis*) – surely the most evocative visual symbol of the savannah in East Africa. Here also are dense thickets of other acacia, including whistling thorn and groves of golden-barked yellow fever trees.

The lands over which the wildebeest move contain more than plains and woodland – there are also shallow alkaline lakes studded with flocks of rose-pink flamingoes, deep freshwater rivers and rivulets, and the dramatic Olduvai Gorge, created by water, that carves its way through the southern Serengeti. Olduvai is full of hominid fossils bearing witness to the emergence, gradual rise, and eventual dominance of the hominid species. Extraordinarily, 3.5 million years ago, our distant ancestors fought to survive in a landscape very similar to that we see today and among animals like guinea fowl, hares, and giraffes that have also changed very little. Others have long since gone, leaving just tantalizing clues to their existence. And some, like the wildebeest, with remains dating back a mere 1.5 million years, are newcomers to the scene.

Wildebeest can be both highly gregarious and mobile or live in small, sedentary herds. Broad muzzles and lips allow them to feed, and quickly,

Moonshine (left): seen through the branches of an acacia tree in the Ndutu woodlands, the moon glows bright in the early evening and, as the calls of the hyaenas echo through the still air, the woodlands take on an air of mystery and intrigue.

on short grasses. Possessors of great stamina, wildebeest are built to migrate, with long, slender legs and flexible backbones. They're not the only migratory antelope in the world but, sadly, over recent years, increased pressure from rising human populations has stopped many other migrations in their tracks. Nor are wildebeest immune from disturbance. Once a subspecies of wildebeest migrated to the very edge of Nairobi, but no more.

So what exactly is the migration?

This is a question that often puzzles those who come here on safari. Surrounded by hundreds of thousands of wildebeest spread out across the plains in golden sunlight, their question is, "Where is the migration?"

The answer lies all around. Even when they're apparently peacefully grazing, the herds are on the move almost constantly. It's only occasionally that wildebeest bunch together in numbers that beggar belief or struggle across a river, like the Mara that bisects the northern Serengeti, or the Grumeti that winds its way through the western plains. These are the spectacular river crossings shown on television, when thousands of wildebeest are forced to brave the onslaught of hungry crocodiles. Such events take their toll on wildebeest, with scientists estimating that migration increases mortality by some 3 per cent. To counteract this, of course, the benefits have to outweigh the disadvantages. And migration means that larger populations can accumulate, since moving avoids over-exploitation of resources. Generally, though, the wildebeest's lives are relatively sedate. When there's enough food and water, they disperse, spreading out over the plains or through woodland; and even though their numbers are massive, they are dwarfed by the huge landscapes.

May or June is another date in the wildebeest calendar that is synonymous with spectacle. This is when the wildebeest mate and it's called the rut. Prime bulls establish small territories within the herd. Defending their patch of land vigorously, they try simultaneously to chase other males away and to stop females from leaving. The result seems to be pandemonium as bulls bellow, stamp, clash horns, kneel, horn vegetation, and roll while females call for calves lost in the frenzy.

Spotted hyaena
Crocuta crocuta

Spotted hyaenas have evolved a unique lifestyle to enable them to keep in touch with the herds. Leaving their cubs behind in dens for up to five days, individuals regularly go on commuting trips to bring back food.

They also have one of the most complex social lives of any mammal, with females dominant over males and one female at the top of the social ladder. Attempted coups are not unknown and, during the struggle to usurp the alpha female, sisters can turn on sisters, swap sides, and then swap back again. In order to survive, the top female has to maintain a tight-knit web of alliances and allegiances that it's not unduly anthropomorphic to describe as "Machiavellian"!

KEY FACTS

Movement can travel 140 km (87 miles) in one round trip and spend 46–62 per cent of the year on the move.
Cubs an average of two per litter, but up to four has been recorded.
Parenting cubs accompany their mothers to learn the major routes.

Sophisticated society (below): long-running research projects in the Serengeti have revealed that hyaenas rival humans in their complex system of allegiances and alliances.

Wildebeest create spectacular displays for the human observer but, more important than this, they play a vital role in the whole ecology of the area. First the carnivores. It is largely thanks to the numbers of these antelope that an estimated 3000 lions and 9000 spotted hyaenas thrive here – the highest density of these species found anywhere in Africa.

Second, the herbivores. As the wildebeest munch and move across the plains, they're doing more than keeping themselves alive. They are creating ideal conditions for other animals to flourish. Thomson's gazelles need shorter grasses for food, so fare better as a horde of wildebeest takes the edge off the longer, tougher grasses. Territorial male wildebeest vigorously horn small trees and bushes on the edge of the woodland, keeping the rapid growth of saplings in check and leaving space for grasses to flourish.

As long as there is enough rainfall, the large numbers of wildebeest also encourage more vigorous and nutritious grasses. Their urine and dung increase levels of sodium and nitrogen, which promote healthy grasses. In addition, grazing leads to an increase in nitrogen in the leaves, speeding up the decomposition of leaf litter. That means energy flows through the system fast, supporting the wildebeest and all the other herbivores around.

Lofty lookout (above): cheetahs use trees and termite mounds as vantage points to scan woodlands and grasslands for prey or enemies, but they are somewhat uneasy, clumsy climbers.

Impala (above): some of the most beautiful antelopes in the world – and the most watchful, with ears and eyes tuned to every sound and movement around them.

Immediately to the north of Lengai and the crater highlands lie the short-grass plains. It's a small area where, at times, the herbivore density peaks at over 238 per sq km (616 per square mile) and these attract the greatest diversity of carnivores anywhere in the world. In addition to the lions and spotted hyaenas, there are striped hyaenas, cheetahs, leopards, and three species of jackal, with a host of scavengers ranging from vultures to dung beetles.

With some three million herbivores and thousands of carnivores gathered in the relatively small space of the Serengeti ecosystem, this is the largest concentration of big mammals surviving on Earth today. But take the wildebeest away and the huge Serengeti landscapes would be a pale shadow of their former self. For 50 years, television documentaries, feature films, books, and magazines have kept the profile of both the Serengeti and the wildebeest high. As a result, tourist numbers have increased and the travel industry has become one of the top three foreign exchange earners for both Kenya and Tanzania. This means that today, despite ongoing pressure from the spectral trio of politics, poaching, and the burgeoning human population, enough land has been set aside as conservation areas to let the migration continue almost without hindrance, so wildebeest numbers have remained high. The most recent census, in 2006, recorded some 1.073 million animals – good news for us, since six months later we made our first foray into the Serengeti to film for *Nature's Great Events*.

Zebra
Equus burchelli

Zebra are one of the most adaptable and successful grazers in the ecosystem, often eating the long, tough stems of grasses as well as feeding on the shorter, more succulent growth long after the wildebeest herds have moved on.

Although they migrate following much the same patterns, zebra move in shorter, more circular movements than the wildebeest, though these mini-migrations are still dominated by rainfall promoting the growth of grasses. Zebra live in small herds consisting of a stallion controlling a so-called harem of five to six mares and their foals. Generally the stallion remains in control until he dies or becomes weak through disease, old age, or injury.

KEY FACTS

Height	up to 140cm (55in) at the shoulder.
Weight	up to 250kg (550lb).
Foals	mares can give birth at any time of year.

"Wildebeeest abstract" (next page): crammed together, shoulder to shoulder, in the rain on the shores of Lake Masek. This artistic image shows form and composition briefly transcending individuality.

Weathering the storm (right above): zebras bunch together, backs to the wind and driving rain. This is the moment that some lions use to sneak up, unseen and unheard.

The lion's share (right below): male lions generally dominate the scene of a kill, taking precedence over females and cubs, and eating their fill before the others.

Pride in peril
Amanda Barrett

Owen Newman and I have worked together making films for the BBC for over 20 years, with most of them focusing on the animals and parks of East Africa. With *Nature's Great Events* we had an unrivalled opportunity to captivate people with the power and beauty of the spectacle of the wildebeest migration – something that we had longed to do for years.

Right at the outset, we were aware we had a problem – how could we shed new light on one of the most photographed events and national parks in the world? After a period of intense soul-searching with the Natural History Unit, they agreed to let us focus on one small area in order to watch the wildebeest leave and return later in the year. In this way, their arrival would be the event and a fitting climax to the film – a refreshing antidote to the approach of following their migration that, we felt, had been done before.

In our view, there was only one place for us to be based – the area around Ndutu Safari Lodge in the south of the Serengeti. The area is a gem. Straddling the borders of Serengeti National Park and Ngorongoro Conservation Area, it contains a great representative sample of the habitats within the Serengeti. There are two seasonal lakes, known as Ndutu and Masek. There are acres of rolling acacia woodland interrupted with vast stretches of grassy plains. And if the rains are good, the mineral-rich grasses attract huge herds of wildebeest from November through to April or May. But Ndutu in the dry season has very little fresh water, so once the wildebeest have gone it can also epitomize the dual problems facing resident animals in the Serengeti – no water and no wildebeest.

The next question was which animal could we film that would show the wildebeest's return as a truly great event? Wildebeest have the greatest impact on predators and, out of all them, it is the lions that find catching enough prey in the dry season most challenging and for whom the return of the wildebeest could be a life-changing experience. We knew there were three prides living around Ndutu, but which should we choose? Like any decision, it wasn't one to take lightly; with the time available for filming, we had just one chance to get it right.

One of our lead characters (right): a female from the main pride featured in the programme surveys the world around her while the wildebeest make a hurried exit.

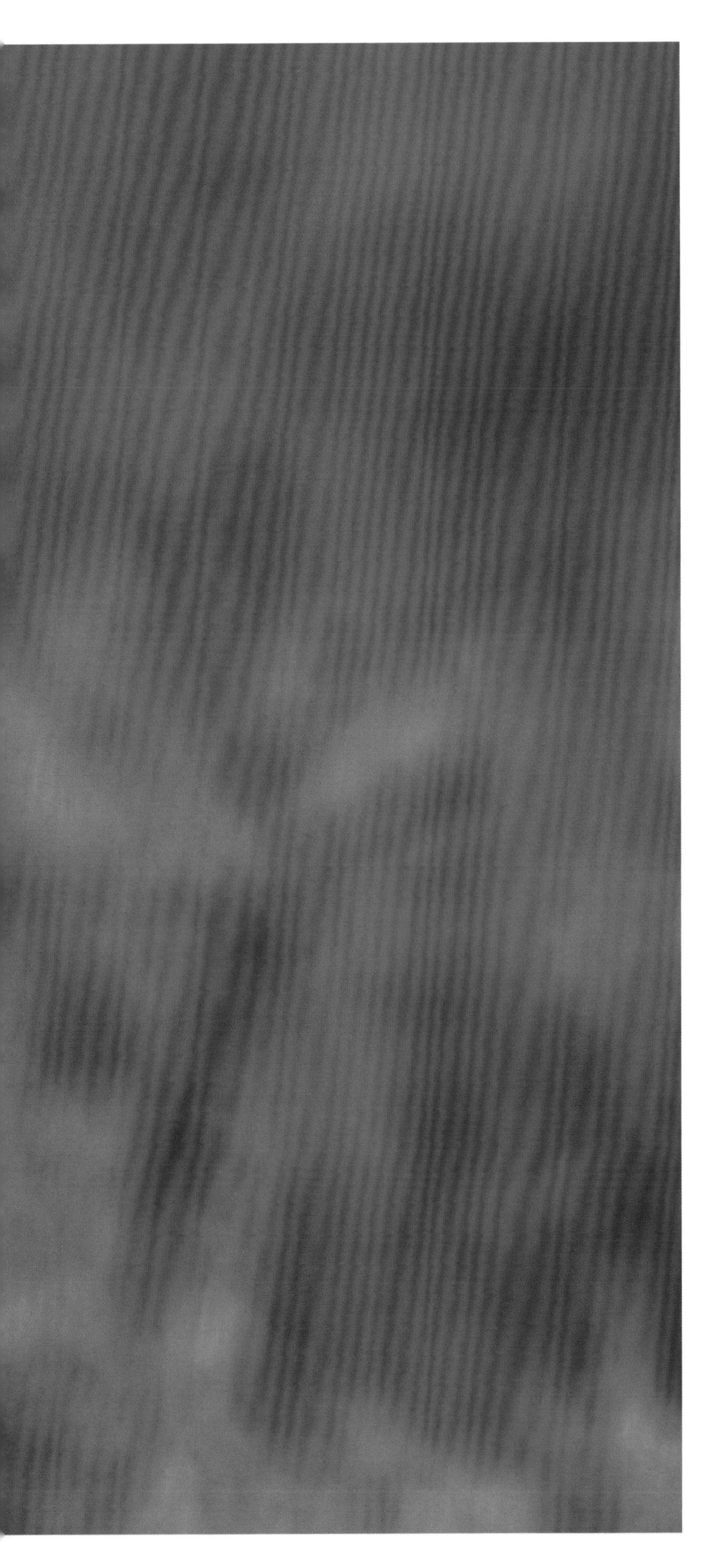

I often feel that getting the footage for a wildlife documentary is just like gambling. There are two wild cards – the animals and the weather – and, at a certain point, the money runs out. We had 12–14 months in to which to fit our seven or eight months of filming before the BBC schedule demanded that editing start. So when should we go? Starting with a short trip in mid-June to focus on the wildebeest rut and choose our lion pride seemed the best bet.

After that, we settled on a longer period to film the peak of the dry season (or so we hoped) from early August into November. And finally, we reckoned we'd be able to catch the best of the rains in January, February, and March, with an extension into April if the budget allowed. Prior to filming, we spent several months discussing, drafting, and rewriting scripts to ensure that key storyline points were covered, but there was still an

Lion
Panthera leo

Muscular and lithe, a single lion is so strong that it can bring down prey twice its own weight, with wildebeest and zebras being particular favourites. Generally, lions live together in a pride and occupy hunting ranges large enough to support them during the driest of dry seasons; however, females often go off to spend time on their own or in a closely related pair, so it's quite rare to see an entire pride together.

Adult females are normally related, with generation after generation living in the same area. If pride numbers fall due to disease or injury, sub-adult immigrants are occasionally allowed in. Young males are usually thrown out of the pride when they are about two or three years old, and after a year or two wandering the Serengeti they may link up with others to form coalitions that subsequently take over other prides.

KEY FACTS
Height up to 120 cm (47in) at the shoulder.
Weight up to 189kg (416lb).
Density up to 3–4 adults in 200sq km (77 sq. miles).

Photo ID (left): lions can be identified by the top row of their whisker spots, each of which is as unique to an individual lion as fingerprints are to us.

unknown factor: what animal dramas would we find to illustrate them? I knew from experience that, for any film maker, however well prepared, the natural world has plenty of tricks up its sleeve.

The curtain rises

We didn't have long to wait and it was more a shock than a surprise. It also coincided with our first glimpse of a lion pride and, oddly enough, it was thanks to a plant. The wildebeest had left Ndutu a few weeks before and, despite heavy rains from January to April, the ground was drying out fast. But, unusually, there were huge, dense swathes of one particular plant, cordifolia, still standing tall. Encouraged by the rain it had romped away in a cavalier fashion and now eclipsed the rest of Ndutu's vegetation. Growing about 1.5–1.8m (5–6ft) tall, it blanketed most of the woodland and much of the grassy plains and, as we were about to find out, it completely obscured the animals.

Although Owen was near the lions all day, the cordifolia was so dense that he could see only a few tails and ears every now and then so filming was out of the question. Then, just before dark, the lions emerged into a tiny clearing of grass. There were four adult females and ten cubs. It was an unusually high number and though they were only nine months old, one or two were already thin. Unfortunately, though, it was now the beginning of July, our time had run out, and we had to wait for the next trip and the dry season.

Owen was back in August. I find the dry season in the southern Serengeti is a forbidding world. As soon as the sun rises, the breeze begins. By midday, strong hot winds are blowing across the plains and, as the ground heats up during the day, tall, narrow dust devils twist high into the air as they undulate across the horizon. With no more rain expected until October or November, combined with scorching temperatures during the day, the dry stands of cordifolia also created a major fire hazard.

Moving in for the kill 1 (above): at the end of a tiring chase, a cheetah has to tackle a Thomson's gazelle with great care, since the horns can cause a lethal injury.

The wildebeest and zebra had long gone, followed by most of the hyaenas. A few Thomson's and Grant's gazelles remained out on the plains, though if the year turned out to be ultra-dry, the majority of the tommies would head for greener pastures in the scattered woodlands. Herds of impala were among the acacia, nibbling on fallen seed pods, so cheetahs and leopards were well provided for. In contrast, lions, who are not as fleet of foot as cheetahs or as stealthy as leopards, faced an uphill struggle, as Owen found out when he caught up with our pride a couple of weeks after his arrival.

The four females looked fit and well, but it was a different story for the cubs. There were only seven left from the original ten and, very sadly, all were emaciated. Some could barely walk but they knew that if they got left behind it would be the end of the road. They had to keep up with the adults who, in turn, couldn't stop or wait since they had to find a good place to hunt. As Owen filmed, one cub gave up the struggle and lay down to die right out in the open. Shockingly, he was just skin and bones.

Some 5km (3 miles) further on and in the middle of the morning, when the land was quivering in the heat, the females caught a warthog. Never had we seen so many lions – four adults and six big cubs – feeding on such a small meal. There was no time for any of them to fight over the meagre pickings; instead each hunkered down and grabbed what it could.

Next day, despite our best efforts, we couldn't find the pride. We searched for days, fighting the cordifolia, looking under trees, gazing out across the plains, scanning the edges of the lakes, and – a sure sign that lions are on a kill – looking for vultures circling in the air. It was all to no avail. The problem is that lionesses have a big and relatively fluid home range that expands and contracts dependent on the availability of prey. We had no idea where they could be. Also, if lions are lying flat on their side, as they do on average 23 hours out of 24, you can drive close by without noticing them. Time was trickling away. Luckily, though, something totally unexpected was brewing that would have a huge effect on the film.

Cheetah
Acinonyx jubatus

With a streamlined frame, these spotted cats are built for speed and hunt mostly during the day. Sub-adult males often form coalitions with their brothers or with other solo males in order to hold territories through which the more nomadic females travel in search of the mobile herds of Thomson's gazelles, their favoured prey.

In order to become the fastest mammal on land, cheetahs have sacrificed the strength and weapons needed to defend kills: it's estimated that cheetahs in the Serengeti lose some 10 per cent of their kills to other predators, generally spotted hyaenas and lions. Cheetahs give birth in thick cover but by five and a half weeks or so the cubs are able to follow their mother wherever she goes

KEY FACTS

Lifestyle females live alone or with their cubs.
Offspring cubs are dependent on their mothers for 18 months or so.
Hunting cubs start to practise catching and killing between 9 and 12 months.

Tale of the unexpected

For a month or so, a series of earth tremors had been rocking the area and shaking our cars. Then there was a shimmering mirage of hot air, shaped just like a genie, hanging over the summit of Ol Doinyo Lengai. Everyone knew something was going to happen, but what and when? The answer came on 4 September, when the mountain erupted, producing a thick plume of steel-grey ash that billowed in mesmerizing clouds from the summit.

Moving in for the kill 2 (right): sadly for them, warthogs are the main prey for Serengeti lions in the dry season, when buffalo or giraffe are absent or hard to find.

Fighting to stay alive (above): we followed the struggles of this female cub with bated breath as she battled to keep up with her pride during the dry season.

Owen and I had filmed inside the crater for others of our programmes and had found a monochrome landscape of white ash and black lava. Smoke gently wafted from tiny cracks around the rim, and bubbling black lava oozed from perfectly round domes on the crater floor. It was an alien world and, like many other people, I was caught in its spell.

As long as the eruption continued, we had an unbeatable chance to film something truly epic. But we had to get up into the air. For this, we needed permission from the Tanzanian authorities and, while we waited, we monitored Lengai. On most mornings, we were able to catch a glimpse of the summit with ash still gently billowing before it was lost in haze. Caused by winds whipping up huge quantities of dust and smoke from dry-season fires, the air fills with smog and makes it hard to see. But as we peered anxiously through the haze and the days passed, the ash clouds grew smaller and we began to worry that we'd lost our opportunity.

There was a silver lining, though, in the cloud. Although the wait was nail-biting, we had an opportunity to look for the lions again. At last Owen found the four females. Tragically they now had only two cubs with them. Both were male and one, in particular, was pitifully thin. During the middle of the afternoon, one lioness started calling and searching for her missing offspring. All the other lions joined in calling, even the thinnest cub. But there was no reply and, after an hour or so, she lost heart. As the light went, the whole group got up and wandered westwards. Had the other eight cubs died or would the pride find some or all of them? There were no answers over the next few days. We simply couldn't find any sign of the lions. But just as I was being driven to despair, the permits arrived. We should have guessed that this wasn't the end of the story, though. Right at this moment, the weather closed in and with a vengeance. Visibility was so poor that Owen and the pilot, Angus Simpson, agreed that it wasn't worth taking off. We couldn't film Lengai – and the lions had vanished into thin air. It seemed as though the gods were conspiring against us.

Tragedy

While waiting for the clouds to clear, we had to find those lions somehow. In mid-September our persistence paid off when, three days after the mother had been calling for the lost cubs, Owen found one. It was a little female and she was all on her own. Although very thin, she was just about strong enough to keep calling, walking, and searching for her family.

It takes at least two years for a female lion to learn to hunt successfully enough to live on her own – and even then most females prefer to stay for the majority of time with their natal pride and to hunt in a group with their sisters. At 10 months old, this one had a long way to go. If she didn't find her mother and aunts, she didn't stand a chance of survival. She knew it, so she kept on calling, hour after hour.

Heart-rending though it was, there was nothing that Owen could do but stay at a distance and wish her luck. His spirits rose late morning as she pricked her ears, eagerly broke into a trot and disappeared behind a bush. He hoped she'd found the pride, but it was only another lost cub – a male – and in a much worse condition than she was. Barely able to raise his head and without the strength to call, his eyes were nevertheless bright with the will to live. It was obvious, though, that he was going to die. The female kept on calling for the pride, but there was still no reply.

Owen found it hard enough to watch, let alone film and by mid-afternoon, he felt he should leave them. He went back the next day but there wasn't a sign of either of them. It was a very long 14 days before we found out what happened next.

Meanwhile, our patience with the volcano was rewarded. The clouds lifted and Owen and Angus were able to take to the air with the camera strapped into prime position on the wing of a small Cessna. It took over 45 minutes to fly from our base to Lengai. As they got closer, the haze masking the volcano from view parted a little and they were excited to see a faint wisp of ash. Closer still, and the clouds of grey ash were thicker and more defined. A strong south-east wind was blowing the ash over the Salei Plains and curtains of ash hung in the air over the grasslands below. At the summit was the grey ash plume forcing its way up and through the bright white cumulus clouds. Owen said that to finally get so close, and to get some footage, made it one of the best days of his life.

And our luck held. The ash continued to billow, and over the next few days Owen and Angus flew to Lengai several times trying to capture this unique event in a variety of different ways and under different weather

Ol Doinyo Lengai

The Masai call this volcano "Mountain of God" and it's certainly an awesome landmark in a spectacular landscape. Looking like a child's drawing of a perfect conical volcano, it rears up from the Great Rift Valley floor to an altitude of 2740m (8987ft) above sea level. In Tanzania, the walls of the Northern Rift are studded with huge volcanoes that are mostly inactive now – Lengai reminds us of what once was. Its steep sides have been eroded by rainwater run-off into deep fissures and canyons that, from a distance, form beautiful abstract patterns of light and shade. At the top, there's an active crater divided from an inactive one by a high, sheer wall.

Lengai is a typical stratovolcano, formed from layers of lava and pyroclasts with a central vent. Its last major eruption was in 1966, but compared to the huge volcanic eruptions in Indonesia and the Andes that was a small though impressive affair. Since then, there has been minor activity, including explosions of pale grey ash from January to March 1983 and August to November 2007. Between those dates, there has been much excitement with black lava bubbling, oozing, and trickling in the active crater. Emerging at a temperature of 600°C (over 1100°F), the black lava is cool compared with that of other volcanoes. It is also rich in sodium carbonate, which turns greyish white as it cools, giving the top of the mountain an unusual "dirty snow" appearance.

In addition, the active crater floor has changed a lot over the last 20 years or so. Rounded, hollow domes of cool lava and ash cones as high as 7m (23ft) or more have risen and then collapsed. At night, some of these ash cones glow red in an eerie, otherworldly way. Steam jets smoking on the rim are the result of rainwater trickling down through cracks in the rock and turning to steam as it reaches a heat source. Then it expands and, since it's under pressure, it's forced upwards again, escaping through small cracks in the surface.

After the explosion (above): the crater of Ol Doinyo Lengai showing the cavernous hole left behind by the eruptions.

conditions. During the last trip, Owen dreamed up a daring plan – if we could hire a helicopter and if the eruption continued, maybe we could land on the top of the volcano to get even better pictures.

Never in my wildest dreams did I imagine that this would be possible. But that's what happened in late November. Our helicopter pilot, Mike Watson, landed right on the top of the small pinnacle that divides the two craters. He switched the engine off and out we got. Below our feet, the crater wall fell smoothly away and we were able to look right down into

Working on a knife edge (left): Owen Newman films an ash eruption of Ol Doinyo Lengai from the summit point between the inactive and active craters.

the hole. We gazed in awe at the clouds of black ash as they swirled around before gathering their strength and billowing out. It was completely transfixing. There wasn't a sound and, amazingly, even at the lofty height of 3000m (just under 10,000ft), not the slightest hint of a breeze.

And that, I felt, was lucky, since the pinnacle was so small that the tail of the helicopter jutted right out over the edge of the volcano. There was nothing between it and the plains far below. I didn't want any of us to put a foot wrong and I hoped the helicopter engine would start up when we needed it. Spending the night up there wasn't an experience I could imagine with any relish. Part of me expected the volcano to go bang at any moment. I had never been in a place so primitive, so truly wild, and so far removed from the sphere of human influence. It brought home to me that our presence – both as individuals and as a species – was, in geological time, completely insignificant. A feeling that was both deeply unsettling yet, for me, strangely uplifting. As a dark curtain of rain raced across the plains to the east of us and we looked down at the rainbow that suddenly

Ol Doinyo Lengai (above): a potent but spectacular mix of toxic charcoal-grey ash and apparently benign white fluffy cumulus clouds.

Skin and bones (above): we had never seen such emaciated cubs as these, nor been forced into the position of impotent spectators during such an epic struggle for survival.

materialized, I thought that we must be the luckiest people alive. Equally, as we took off, I was hugely relieved that we had all come out of it in one piece.

Our curiosity about the plight of the cubs was finally resolved – and happily – in between flights to Lengai. We found the four lionesses with two cubs. One was the little female: her resolve to live had carried her through. But although the second cub was a male, it wasn't the one we'd seen her with two weeks earlier, so her erstwhile companion had almost certainly died. From June to September we had seen the pride lose eight youngsters, with the last five dying in just four or five weeks. It was a poignant example of how tough life can be for lions in the Serengeti's dry season.

Transformation

The lion story plus the volcanic eruption were already a potent mix, but there was another surprise in store for us – a major bush fire that rushed through the woodland and encircled Ndutu Safari Lodge. The owner had the foresight early in the dry season to create a firebreak around the lodge, so it was relatively safe. However, there was no stopping the fire in the rest of the area. For 36 hours, flames roared through the woodlands and raced across the plains before they finally burned themselves out. Owen focused on filming the fire at night when the lurid reds and oranges looked their most dramatic. Being so close to a leaping wall of heat made his pulse race, and the overwhelming roar of dry vegetation bursting into flame was almost deafening and very frightening. But, like the other guests as Ndutu, he was lucky: if the worst came to the worst, he could get into a car and drive away. Tragically, hundreds of animals, reptiles, and birds couldn't.

Tough challenges (left): prey can be hard for lions to come by during the dry season. Here a male drags a dead cow onto the shores of Lake Masek in the southern Serengeti.

Conflagration (next page): a bush fire rages through woodland at Ndutu late in the dry season. The flames are so tall that buffalo weaver nests have caught alight in the tops of the acacia trees.

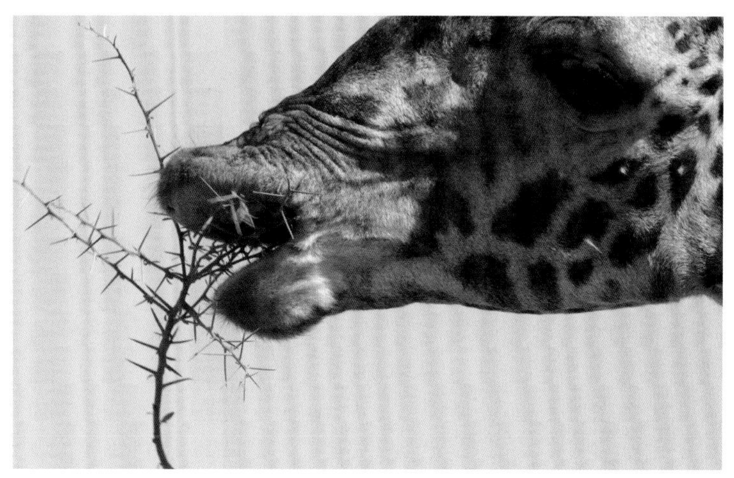

Height advantage (above): although flames had scorched the lower branches, giraffes could still browse fresh growth from the tops of the trees that hadn't been killed by the fire.

Habitués of dense undergrowth (above): these caracals seemed confused by the sudden lack of vegetation and, unusually, became easy to see.

In the morning, smoke curled up into the air from burning wood and the ground was covered in black ash. I found it deeply unsettling to find the charred remains of chameleons, ground-roosting birds, tortoises, and many others that had been burned alive. The vast swathes of cordifolia that had hidden the animals from our searching binoculars were just a memory and, extraordinarily, Owen saw both leopard and caracal inside half an hour. Experts at hiding, these cats are rarely seen but, suddenly, their cover had been blown. The change was so profound that many animals seemed dazed and confused in the aftermath.

According to the theory, fire is a necessary part of the savannah ecosystem, keeping the woodland in check and allowing grasses to flourish. But in practice this fire – set by Masai or poachers – came so late in the season and the tall, thick cordifolia was so desiccated that the

The morning after (left): the big fire has transformed the landscape. Animals such as elephants that could flee the flames were fine; hundreds of others were not so fortunate.

An easy meal (above): many birds, such as hoopoes, struck lucky after the fire, with plenty of insects in plain view.

Sudden exposure (above): steenbok are antelope that rely on thick cover in which to hide; suddenly their lives were turned upside down.

A poignant farewell (above): this was our last sight of the thin pride. We had no idea whether the cubs would survive.

results seemed disastrous. Two weeks later, I was depressed to see how many huge acacia trees in their prime had been toppled by hot cinders lodging in cracks in their bark and slowly smouldering through the trunk. As I moved through the woodlands, it looked as if a careless giant had been on the rampage with a chain saw. In the light of such devastation, it seems totally churlish to admit that it made life easier for us. But the truth is that, without the cordifolia, we could see for miles and our chances, we thought, of finding the lions improved immensely.

It didn't turn out like that. We spent several days driving through the woodland and along the edge of the plains with no results. We found giraffes and impala nibbling on tiny shoots while elephants munched on bare branches and twigs, breaking them off with a loud crack that resonated through the scorched woodland. Ground hornbills and kori bustards stalked across the plains, keen to grab any hapless insect or reptile that strayed across their path. The biggest concentrations of raptors we'd ever seen here congregated in the woodland – Montague and pallid harriers, Eurasian buzzards, tawny eagles, black-shouldered kites – all indulging in the easy pickings as insects and mice were forced to forage with no cover.

Of all the animals in trouble, it was the steenbok I felt most sorry for. Knee-high antelopes, they're solitary animals that rely on thick vegetation in which to hide. They don't have the speed or stamina of other antelopes that are used to running from predators. Now they had no choice but to crouch, chestnut coats bright in the sun on stark, blackened plains. Since they were right out in the open, they were easy prey for any passing cheetah.

Our time was running out, but just before we had to leave we finally hit a lucky streak with the lions and found them on several consecutive days. It was an almost undreamed-of luxury. Even better, I was amazed to see the two youngsters still hanging on to life, although I was shocked to see how gaunt they were with rough fur and patches of mange and bare, black skin. In stark contrast, the four adults were sleek and plump. We were able to film them hunting warthogs again and on most days they caught at least one. The rains would bring the wildebeest and zebra, and their problems, surely, would be over. But before this we estimated that they still had two or three weeks to go. I had one over-riding question – could the cubs make it?

We had to be patient because, for now, like the lions, we sadly had to husband our resources. Our allotted money for this trip had been used up and we were due to return to England. As we left on the last evening, we didn't know if we'd ever see the pride again.

Taking the change in their stride (left): Leopards like this young male exploited the opportunity to explore their new world.

The rain

Back in England, the lions' plight was never far from our minds, especially as things on the Tanzanian weather front hadn't worked out as we'd hoped. The first rains that we had thought would bring relief to the lions didn't come. Instead, the area got drier and drier, without a wildebeest or zebra to be seen. Then we received two reports that worried us immensely. The first was from a tour driver who had seen a young male lion lying in the gully where we'd first found the pride all those months ago. Apparently, he was so weak from hunger that he couldn't get to his feet. More news, only a few days later, was courtesy of a Dutch filmmaker who'd spotted a hyaena near where we'd last seen the pride, feeding on the skull of a sub-adult lioness.

Then in late December we heard that the rains had finally arrived. From our past experience I knew that Ndutu was being transformed by one of nature's great events. I love the smell as the first spatterings of cold rain hits the hot dry earth. It's so strong it's almost tangible – a warm, febrile scent of promise. And there's not long to wait. Although the rainfall has

been minimal, short green shoots spring up within a day or two, just like stubble on a man's cheek. And even this makes all the difference. As you look into the sun, a green shimmer banishes the tawny silver hues of the dry season. Instead there are vivid emeralds and limes with backlit grasses glowing with an almost fluorescent fervour. The warthogs, steenbok, and other antelopes who've hung on through the parched times are right on the spot to exploit the nutritious growth. And it's a good thing, because an army is on the move.

The first sign of its arrival is often at night. Lying in bed, you are suddenly aware of a faint roar like surf breaking on a distant shore. Amid the clamour, the bell-like calls of a pearl-spotted owlet ring out, clear and bright. A hyaena whoops, a spine-chilling clarion call. Slowly but surely, the roar grows louder. And in the morning, there's a miraculous sight. Where before there had been empty spaces, groups of wildebeest now occupy almost every nook and cranny. They're in the woodlands, out on the plains, and stomping across the lakes in long lines. Every now and then

The wet season (above): colour returns to the parched landscape; the downpours can be sudden and violent, but after months without rain it's a welcome relief.

Elements of the Serengeti
The wet season

Storms
During the rainy season, a small fluffy cloud in the mid-morning can build into a looming thunderhead by mid-afternoon. Lightning strikes out on the short-grass plains have been known to kill giraffe and groups of wildebeest, so the billowing mass of black cloud can be deadly. Influenced by monsoons from Asia, hurricanes from Madagascar, weather patterns from Lake Victoria, and local topography, the first rains start in the north and move south; later on in the year, the rains come from the east.
There are years when rainstorms are so fierce that floods sweep over the grassland; the next year there may be a severe drought.

Mother-of-pearl clouds
Arguably the rarest and most beautiful clouds in the sky, nacreous clouds form in the twilight hours at sunrise or sunset. And they're way up in the stratosphere, more than double the height of the more common clouds. While the rest of the lower cloud formations are in shadow, these clouds are caught by the light of the sun just over the horizon. Up here, the temperature is around -85°C (-121°F). Rarely seen, these magical clouds are also the most destructive, speeding up the depletion of the ozone layer.

Grasses
Easily overlooked, grasses such as *Cynodon dactylon* are super-heroes to cope with extreme dry seasons and heavy grazing pressure. They grow in small mounds that, during the dry season, are stiff and dry, apparently dead. In fact, they store starch reserves underground in rhizomes or bulb-like corms. As the rain falls, trickles of water are dammed up between the hummocks to ensure the roots are able to make the most of minimal showers. They're also able to dress themselves quickly in lots of leaf, allowing them to cope with the grazing pressure exerted by the hordes of hungry wildebeest.

Flowers of the plains
During the rains, flowers appear almost overnight and the plains are transformed. Most vivid of all, the daisy-like yellow blooms of *Hirpicium diffusum* almost obscure the grasses for miles. But there are also the tiny red flowers of *Hibiscus* that are so bright, they almost hurt the eyes, and the showy flowers of various morning glory species such as *Ipomoea longituba*. One of the most endearing things I've ever seen is a Masai warrior, dressed to kill, with a yellow and red *Gloriosa* lily tucked behind his ear.

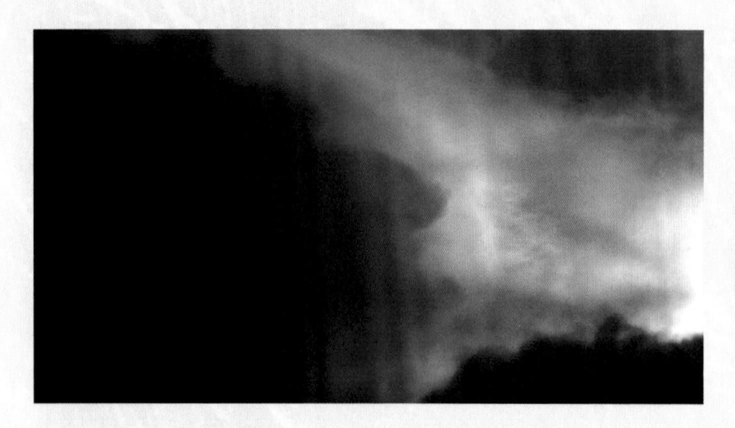

one grunts. It's a low, soft, contact call, but penetrating. And it is this magnified by hundreds of thousands that has created the night-time roar. From now on, the lives of all the animals in Ndutu are going to be very different. I couldn't wait to get out there.

When we arrived in January, we found thousands upon thousands of heavily pregnant females grazing on the plains around Ndutu. The herds dwarfed the resident impala, warthogs, and gazelle that could be seen as tiny stalwart islands amongst the ebb and flow of the wildebeest.

Triumph

Other animals had arrived in force as well. Following the great herds were huge numbers of vultures and spotted hyaenas. Nomadic lions turned up, too, including a female we'd never seen before with four small cubs. The whole of the natural world seemed to be on the move, so where was "our" pride? After hearing the stories of dying and dead sub-adults, we had no high hopes that the cubs would have made it through the dry season. Then, one morning, on the shores of Lake Ndutu close to where he'd first seen them in August, Owen found four lionesses. They were finishing the remains of a wildebeest carcass, surrounded by a group of spotted hyaenas eager to take over the kill. At the heart of the fray were two youngsters, a male and a female – no longer walking skeletons, the signs of mange replaced by thick coats shining with health. Could these really be the same animals? Owen's gut feeling was that they were, but he couldn't yet be sure.

The young male lunged at the hyaenas and, before Owen was able to do much filming, the whole pride disappeared deep into the middle of a sea of scrub and thick bushes, leaving the hyaenas to the kill. Although Owen waited all day, none of the lions re-emerged and he wondered if he'd made a mistake. It seemed so miraculous that the cubs should have survived that he began to doubt his own eyes. He had to find them again. But, as in the dry season, he had to be patient.

Fierce contenders (above): lappet-faced vultures have the heaviest beaks of the African vultures and are the strongest, most intimidating competitors at a kill.

It was days before he caught up with them again. This time, one adult female was missing but he was able to check the whisker-spot patterns of the cubs and confirm once and for all that it was the same pride. The top line of whisker spots on each lion is idiosyncratic, like fingerprints in people, so this is the only sure-fire way of identifying lions. Now that Owen was able to inspect the little female closely, he could also see faint residual traces of mange on her face, where small patches of thinner hair than normal overlaid her black skin.

Although the skies had been grey, no rain had fallen for some time and the fresh grassy shoots were beginning to look the worse for wear, shrivelling in the hot daytime sun. Slowly the plains and woodlands were emptying of wildebeest as they almost imperceptibly drifted towards the west. And then, the clouds gathered and the heavens opened. Rain fell in torrents, day after day. On some afternoons, as the tracks filled with water

A brief respite (above): it was a great comfort to find that the two cubs had not only survived but were relaxed and well fed enough to enjoy playing with each other – and a tortoise. The tortoise survived to see another day!

rushing off the plains, it felt like the end of the world. Then, just as suddenly as they had arrived, the rains stopped again and the sun shone bright on a new world. It was the middle of February and the sky had the clarity of crystal.

Overnight, the wildebeest were back around Ndutu in force. Line upon line of them marched out of the woodlands, leaving behind clear trails of flattened grasses. As I drove out high onto the plains, my heart soared. Wildebeest are among my favourite animals and this was an idyllic scene. Larks and cisticolas sang high in the air above great swathes of wildebeest; family groups of zebras with tiny foals grazed, their stripes pixillating in the strong sun; and small gazelle fawns bounded up from their hiding

places in the vegetation as black-backed jackals prowled through the herds. On days like this, I can't think of anywhere else on earth where life is so abundant, warm, and peaceful. And in the days that followed, every day gave us a different treat: a young aardvark out foraging, unusually, in the daytime; great spotted cuckoos catching caterpillars; a hoopoe singing its heart out; or, like a rainbow gone mad, the rarely seen streaks and feathers of mother-of-pearl clouds high above a dark storm cloud.

Plains of plenty

Some wildebeest had calves already, but the peak of the birthing was yet to come. In the next few days, the antelope gave birth in their thousands. The majority of births seemed to happen in the early morning, so by mid-morning, wobbly-legged calves hugging close to their mother's sides were everywhere. It's a joy to watch and a strategy that pays off. By synchronizing their births to a few short days, wildebeest flood the plains with their babies. Despite the rush of predators, the calves find safety in numbers. They're also the most precocious of all antelopes, able to run fast with their mothers just a few minutes after their arrival into the world.

This is the theory; the reality of the strategy means that during the peak there is frantic activity in all levels of the animal world. Vultures fight over scraps of afterbirth, hotly pursued by jackals. Cheetahs, the swiftest of the predators on the plains, have their pick of the newborn calves. And the noise from the wildebeest increases to deafening levels. Calves call shrilly to their mothers while the females bellow back; it is the only way they can find each other when separated in mad gallops from predators, or during their movements back and forth across lakes and rivers.

Owen finds that filming a wildebeest birth out on the open plains is fraught with difficulty since, understandably, the females are ultra-nervous.

The drought is over (above): a Ruppell's griffon vulture takes to the water. On some days when the air is warm and the sun is bright, the wet season in the Serengeti seems an idyll for every living creature.

He has to avoid upsetting one so much that she runs off as soon as she's finished giving birth. If this happens, the calf will struggle to its feet and imprint on the first moving object it sees – and that would be his car. So Owen was careful to stay at a discreet distance and to find females that seemed calmer than others. The tell-tale sign he was looking for was a tail kinked to one side with a pair of tiny ivory-coloured hooves poking out into thin air.

At this point, the antelope can delay giving birth for some time depending on the danger she perceives – marauding hyaenas are one good reason, cheetahs or jackals another. So, occasionally, you spot a wildebeest walking slowly away, looking for peace and quiet, in what would appear to be a dire stage of her pregnancy. Having filmed several wildebeest births over the years, we still find we're on edge with each one.

If all goes well, she'll finally lie down on her side, her body heaving in steady, rhythmic contractions. Her head turns to one side to watch the calf emerging into the bright new world. Occasionally, she'll get back up on her feet and quickly turn in half-circles, trying to throw the calf out. Speed is of the essence. I once watched an amazing sight. A cow had missed the peak of the birthing and was having her calf on the move in the midst of a vast column of wildebeest on the march through a gully and marsh. She dropped down as the column continued to file past. In just 7 minutes the calf was standing and walking, so mother and baby were able to rejoin the end of the column as it continued on its way.

The big picture

It's impossible to film an event that is as moving as a wildebeest giving birth or the spectacle of the huge herds without thinking, or worrying, about what will happen to them in the next decade or so. In our 20 or so years of filming in East Africa, we have already witnessed huge changes.

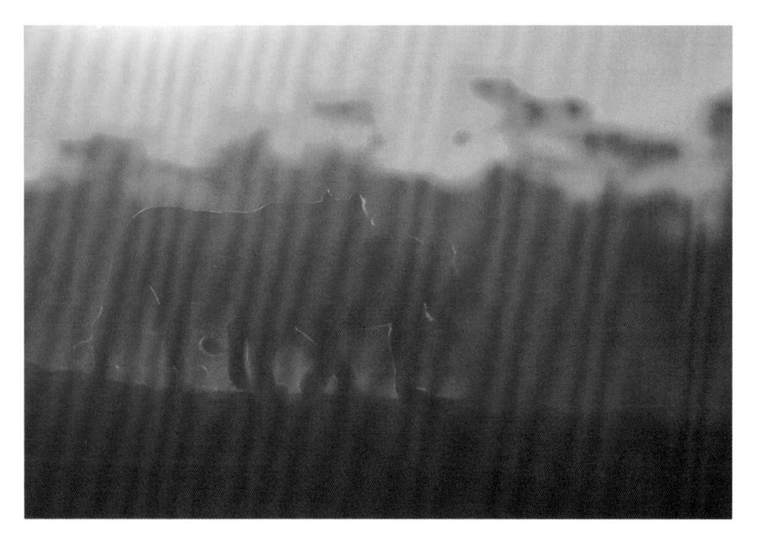

As a result of a massive increase in the human population of Kenya and Tanzania, more and more people are moving into marginal lands on the edge of the parks – or even right into their hearts. Meanwhile, crops such as wheat, grown for cash and export to the West, have replaced the grassy plains in the Mara that were vital for one subspecies of wildebeest, the Loita Hills population. As a result, it is now virtually extinct. Trees inside parks are chopped down for fuel or to be exported as timber to the West or Asia, whose inhabitants are voraciously hungry for exotic hardwoods. Some Tanzanians and Kenyans are growing rich but the vast majority is poor.

In the light of this, both Tanzania and Kenya should be congratulated for having done more than most Western countries to support the principles of conservation – putting aside vast tracts of land for national parks and game reserves. But this can only work for them if the wildlife and the landscape make money – eco-tourism is one answer that's gaining popularity. Sadly, tourism is often far from eco-friendly. Most visitors seem to have no idea that tight schedules and cutting corners result in untrained drivers unintentionally harassing animals to give their guests photographic opportunities. Camp crews leave litter behind and elephants die from eating plastic bags or metal bottle tops. Vast quantities of wood and water are consumed unnecessarily as uneducated guests demand luxury items such as deep baths even in the dry season.

Even worse, our changing climate is predicted to exacerbate the unpredictability of the rains. But despite the pessimism, there is hope that the wildebeest and other animals will continue to go from strength to strength. The Serengeti is the largest and most intact savannah system surviving in Africa today. And each one of us can help keep it this way. If we visit their countries, we can be less-demanding guests. If we don't, we can reduce our ecological or carbon footprints at home. In these ways, the Tanzanians and Kenyans can be supported in their commitment to conservation. A small price to pay to conserve a place where great natural events play out in a landscape where volcanos, such as Ol Doinyo Lengai, still dominate the world around them.

Twist in the tale

In March 2008, Lengai demonstrated that, as in all good stories, it had saved its best for last. It started erupting – and carried on for three weeks. On the most dramatic days, it took just a startling two minutes for thick plumes of ash to power up to over 18,000m (60,000ft) into the air. That's double the height at which many commercial airliners fly. We were even

An uncertain future (left): will the new males leave and let "our" two cubs return to the pride? Only time will tell.

February 2008 (right): the ash plumes from the biggest eruptions of Ol Doinyo Lengai reached an awesome 12–18,000m (40–60,000ft), creating thunder and lightning in the ash clouds.

able to film wildebeest grazing on the Salei Plains in front of curtains of ash. It was such a unbeatable illustration of the birth of a habitat as well as the way it is subsequently utilized that we wouldn't have dreamed of putting it forward as an idea.

Luck was on our side – almost. But we still couldn't track down our pride. Maybe they were finding prey so easy to catch that they were hunting and feeding in the cool of night, and hiding in the shade during the day. But then we finally found them and were proved completely wrong.

Lion finale

We had decided to enlarge our search and look out on the plains in the opposite direction from usual. Amazingly, on the very first morning, there they were. Well, there were three adult females, but the mother and the two cubs were missing. The reason? Two youngish males were with them and mating was well under way. Our guess was that the males were too young to be the fathers of the cubs – in which case, they were a threat to them. Even though he was only 19 months old, the male cub would be considered a rival and killed. The young female, too young to mate with, wouldn't be tolerated and could be badly injured. So we were relieved that the mother was wise enough to keep them out of the way. Cub-killing is an evolutionary constraint brought about by the high number of male lions in the Serengeti. The competition between males for "ownership" of a pride is so intense that most can only maintain their position for two or three years – just long enough to see their cubs mature. With no time to lose, incoming pride males kill their predecessors' cubs to force the females into oestrus quickly, so that they can father their own offspring and have a chance of seeing them into adulthood. It's the only way of passing on their genes.

As time passed, we realized we'd been right. Whenever we found the four adult females and two young adults, the new males weren't with them.

When the males were around, the cubs were missing. One morning, Owen filmed the young male nervously watching at a distance a male with his mother and two aunts. He looked hungry, too, which was particularly ironic: just when food shouldn't have been an issue for the cubs, they were caught between the devil and the deep blue sea. Whenever the females made a kill, the males turned up and the cubs had to retreat fast. Our hope was that the mother would leave her sisters and try to raise the cubs on her own – not so difficult now as there was a lot of potential prey around.

By the end of the month, we thought we had only a few days left to tie up the loose ends in the lions' story and once again the tension was almost unbearable. Sometimes we'd find the lions, sometimes not. More and more

"Flehmen" (left): one of the new males that took over "our" pride, tasting the scent of a female in the air to decide whether she's receptive to mating.

often we'd discover all four females with the males, and no sign of the cubs. The two new males were joined by a third, slightly older male. And a fourth male, much older than the rest, also turned up. Our hoped-for happy ending seemed to be receding.

Then at the last minute we were able to make the budget stretch for longer than we'd originally thought, which was lucky since three weeks later we finally found the cubs again. This time they were on their own without a sign of the adults. Both were sleeping in the sun, looking plump and relaxed. We were perplexed. Had their mother abandoned them? Were they hunting for themselves? There were still wildebeest about and, here and there, a lost calf. We knew it was the best time of year for young and inexperienced lions to try to fend for themselves but we needed proof before we could say, for the film and for certain, that they were OK.

The next day we found the mother – and at a distance the female cub. Both were calling softly, peering at each other across the plain, but there was no way that they could get closer to each other. The mother was accompanied by two males, who were sitting on either side of her staring, with what seemed to be belligerent intent, at the cub. It was stalemate all day, and our feeling that the males were hostile was borne out when the little female got up in the heat of the day and tried to slink into a bush not far away. The minute she moved, both males sat bolt upright and stared at her like cats watching a mouse and, without reaching the shade, she sat down again. Then eventually, the males relaxed. It was obvious that the mother was caught in a Catch-22 situation – she wanted to be with her cubs but she couldn't lose the males. This was borne out in the late afternoon, when the mother sat up and started calling loudly. She continued to call until, in the distance, herds of zebra galloped off with shrill whinnies of distress. It was a signal that a predator was on the move. And, sure enough, a few minutes later, the male cub appeared, a tiny dot on the horizon. But on catching sight of the males he immediately turned tail and ran. Both males galloped after him, stopping only when he vanished from their sight. It was getting dark and, frustratingly, we had to leave.

For a few days after that, we found the female cub alone – then the male cub. They were never together and we didn't see the mother anywhere near them again. Instead, we found her a long way away, socializing with her sisters and the new males. My last sight of the young female cub pulled at my heartstrings. Only 18 months old, she was on her own, lying on her back, a forlorn figure in a vast and empty plain with storm clouds gathering behind her.

After I'd seen her struggle so hard for so long to get this far in life, it was sad to think that I'd never know whether she'd make it or not. Could she survive on her own, or would she be able to join the pride, or her brother again? I tried to take comfort in the fact that at least she was still alive. And I knew it was possible that as the dry season came and the wildebeest left the Ndutu plains the adult females would return to dry season haunts, leaving the males behind, and that the cubs might reunite.

So, surprisingly, the departure of the wildebeest herds might, this time around, actually bring relief to the two cubs. And I hoped that by the time, the vast herds returned in the next wet season, the young male would be old enough to hunt for himself and find a male companion, whilst the young female would be sufficiently attractive to the new males that they'd accept her as part of the pride.

But that's just guesswork on my part and all I know for sure is that we'd filmed more graphically than we'd ever imagined possible; the stark reality behind the scientific facts. For lions, the Serengeti is no Garden of Eden. The dry season exerts a tremendous toll with the statistics showing that 80 per cent of cubs die before they're two years old, from disease, starvation, or infanticide from unrelated males.

I trust we have not filmed our lions suffering in vain. Personally, being involved in making this film means that some of the most harrowing sights and some of the most beautiful will stay with me forever. But more than this, I hope that the Serengeti holds enough surprises to excite the most jaundiced of souls. And that our story of one of nature's great events will touch people in more ways than I can imagine.

A last word (above): as our epoch enters its twenty-first century, I can only hope that the Serengeti ecosystem goes from strength to strength so that vast herds of wildebeest can continue to roam free across some of the most glorious landscapes in the world.

The Great Melt
The Arctic

The polar bear walked towards us. It was a large male, in his prime, his white coat in perfect condition. He approached quickly, taking in our scent, air hissing through his nose. We threw a snowball to scare him away, but it landed short and he kept coming, closing the distance between us with rolling steps of his giant front paws. When he was just 9m (30ft) away we decided to retreat to the safety of the hut. We watched the bear still coming closer, his eyes intent and unafraid. Steinar muttered, "That's as close as you come", and pulled out his flare gun. A pink burst of light shot out and hit the snow just short of the bear, who turned in surprise and ran back to the shoreline, his pigeon-toed back legs skidding on the ice.

In the polar north, the return of the sun, after months of winter darkness, signals the start of a dramatic change. Each year the great melt of the Arctic sea ice opens up vast tracts of rich ocean and coastal habitat, presenting great opportunities for millions of birds, whales, and other migrants that travel from the south to feed and breed in a land where the sun never sets. The ice and snow shrink back to reveal a land of surprising richness and diversity. However this season of prosperity is short lived; the Arctic summer lasts only a few short weeks before the harsh winter world returns again and the resident animals need to draw on all their resourcefulness in order to survive.

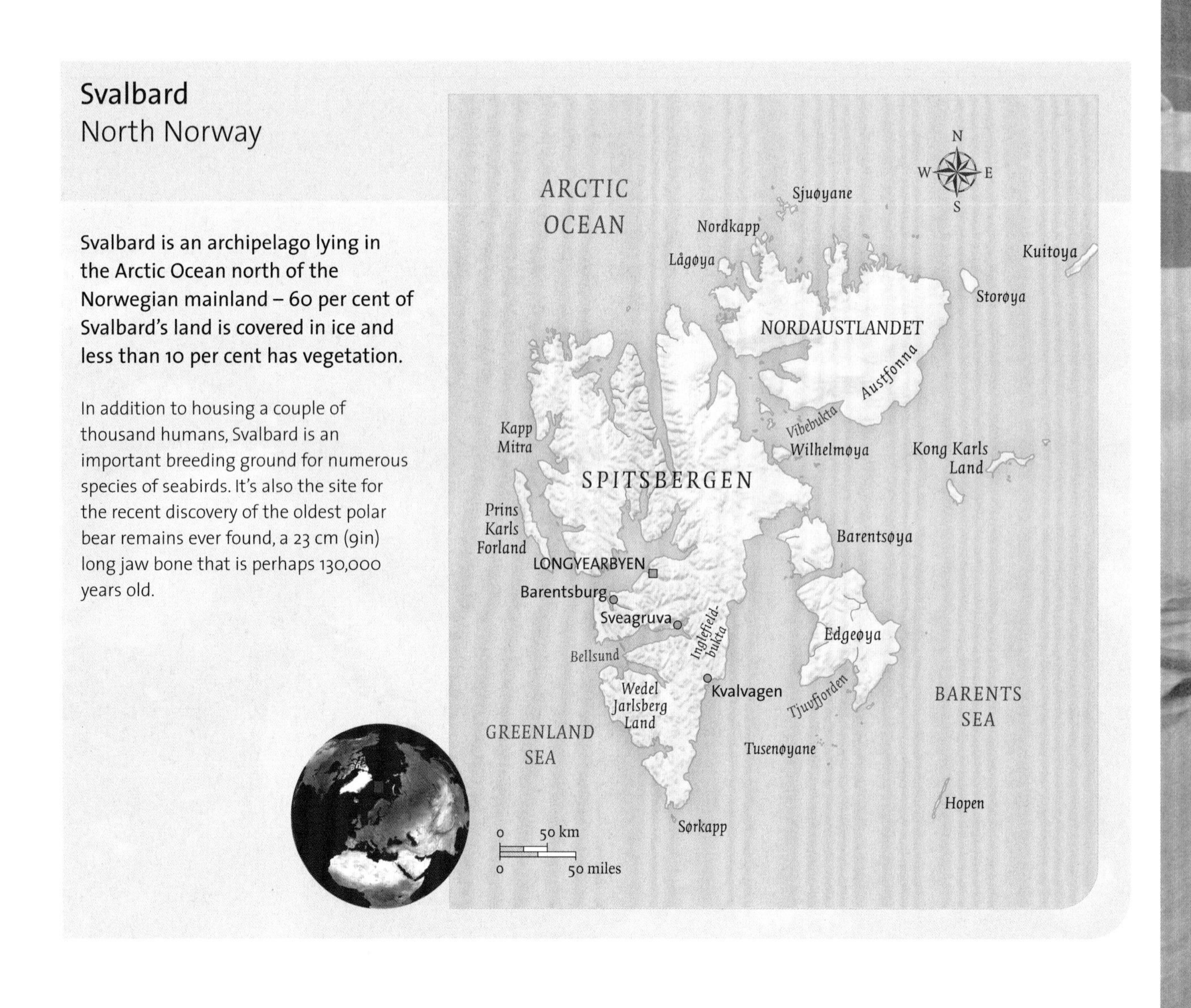

Svalbard
North Norway

Svalbard is an archipelago lying in the Arctic Ocean north of the Norwegian mainland – 60 per cent of Svalbard's land is covered in ice and less than 10 per cent has vegetation.

In addition to housing a couple of thousand humans, Svalbard is an important breeding ground for numerous species of seabirds. It's also the site for the recent discovery of the oldest polar bear remains ever found, a 23 cm (9in) long jaw bone that is perhaps 130,000 years old.

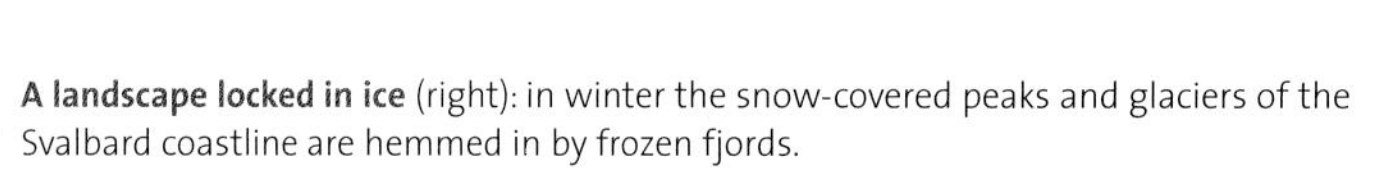

A landscape locked in ice (right): in winter the snow-covered peaks and glaciers of the Svalbard coastline are hemmed in by frozen fjords.

What is "the Arctic"?

Beyond the line of latitude found at 66.56083 degrees north of the equator – the line we call the Arctic Circle – lie 14.5 million sq km (5½ million square miles) of wilderness, the area known as the Arctic. Within its boundaries there are great forests, stony deserts, boggy tundra, pristine mountains, huge glaciers, and permanently frozen ocean. The Arctic, in fact, is best thought of as a frozen sea surrounded by a ring of continents: the north coasts of Russia and North America, and the most northerly outposts of Europe.

With average winter temperatures at the North Pole sitting at a decidedly chilly –34°C (–29°F), it's cold enough for the sea to freeze. Fresh water freezes at 0°C (32°F) but sea water, because of its salt content, won't freeze until the temperature hits –1.8°C (29°F). In the short summer, the North Pole's average temperature is only 0°C (32°F), not enough ever to completely melt the sea ice there. This has led to the Arctic Ocean being permanently frozen over – a vast area of 7–9 million sq km (2¾–3½ million square miles) known as the "permanent Arctic pack ice". In the winter, as temperatures plummet, this area doubles. The sea ice grows until it covers a staggering 14–16 million sq km (5½–6 million square miles). In the summer, as temperatures rise, an average of 7 million sq km (2¾ million square miles) of sea ice melts away.

Snow falls all across the Arctic but precipitation, including snow, is relatively low (about 200mm/8in annually in Longyearbyen on Norwegian Svalbard). The Arctic is generally a much warmer place than Antarctica, where the mean annual temperature is –50°C (–58°F) as compared to –18°C (0°F) at the North Pole. Because of this the Arctic has a much smaller area of permanent land ice (glaciers and ice caps). While 14 million sq km (5½ million square miles) of ice cover Antarctica, the largest Arctic ice cap, on Greenland, measures a mere 1.7 million sq km (660,000 square miles) Other smaller ice caps are found on Ellesmere Island in northern Canada, and in northern Svalbard.

This lack of permanent land ice means the lands north of the Arctic Circle are home to a variety of habitats. In the southern Arctic latitudes, great areas of boreal forest give way to the "barren ground", a huge belt of Arctic tundra that encircles the globe. Here the soil is permanently frozen to a depth of 25–90cm (10–35in). The frozen layer is called permafrost and it is impossible for trees to grow roots here, so the dominant plants are mosses, heath shrubs, and lichens. In the summer the top layer of the permafrost melts, turning the landscape into a boggy network of ponds and marshes.

Further north lie huge expanses of desert. With little precipitation and temperatures cold enough to fracture rocks, these are barren and inhospitable areas where only the hardiest plants and animals can survive.

Ice bear (left): the Arctic is home to 25,000 polar bears, 3000 of which inhabit the frozen coastline of Svalbard.

Iceberg drifting in the summer ocean (above): the blue colour is due to the air having been compressed out of the ice by layers of snowfall when the ice was still part of a glacier.

Sunlight hours

The Arctic receives as much sunlight in a year as the tropics, only it comes all at once. Close to the pole, the sun never sets for almost six months.

POLAR NIGHT | TWILIGHT | NIGHT SUN

JANUARY | FEBRUARY | MARCH | APRIL | MAY | JUNE | JULY | AUGUST | SEPTEMBER | OCTOBER | NOVEMBER | DECEMBER

Why does the ice melt?

The Arctic Circle marks the southern extremity of the polar day and polar night; if you lived right here you would experience one day each year of 24-hour sunlight (on the summer solstice, 20–21 June) and one of 24-hour darkness (on the winter solstice, 20–21 December). Further north the relative periods of "midnight sun" and permanent winter darkness increase until, at the North Pole, they reach an extreme. Here, at the planet's most northerly point, the sun rises on 20–21 March (the vernal equinox) and takes three months (until the solstice) to reach its highest point in the sky. It then begins a slow three-month sunset before finally dipping below the horizon on 22–23 September (the autumnal equinox). This means that at the North Pole there are approximately six months when the sun literally never sets, followed by six months when it never rises. This extraordinary effect is caused by the Earth's axial tilt and its revolution around the sun. During the summer the North Pole is always facing the sun's rays; during the winter it faces away.

These fluctuations in the presence of the sun are responsible for the Arctic's extreme seasonal differences. The lowest winter temperature in Longyearbyen, which is 78 degrees north, is recorded at –44°C (–47°F), whilst the highest summer temperature is 21°C (70°F). Only in Antarctica and the Gobi Desert can higher differences in seasonal extremes be seen. This huge seasonal change in temperature is the reason the Arctic ice melts.

Meltwater pools (above): the sea ice begins to melt as the sun beats down on it, forming pools of turquoise water. These pools absorb more of the sun's heat than the ice itself, causing the rate of melt to speed up.

Before the melt

At the height of the Arctic winter, when there is no sun, the land and sea are locked down and frozen. The shimmering lights of the *aurora borealis* ripple across the sky. This is the land of the midday moon when, for part of each month, it never falls below the horizon.

Compared to the planet's other habitats, the Arctic in winter is devoid of life; only a few hardy animals can survive. On the sea ice, ringed and bearded seals haul out at polynas (large openings in the ice that remain unfrozen even in the winter). They are hunted by polar bears, whose kills may be scavenged by Arctic foxes. On land, lemmings shelter under the snow in nests of grass, wolves wander the barren tundra, and herds of musk oxen break their way through the icy crust of snow to nibble on meagre grasses and herbs. On my first ever trip to the Arctic I travelled to Banks Island in Canada. Animals survive here in relatively large numbers for such a northerly habitat, but even so, days would go by when we would spot only a few distant musk oxen or an Arctic hare nibbling in their tracks. I remember one day we encountered a pack of three pure-white Arctic wolves. They approached us, hugely curious, before heading off into the endless wide open. We tried to follow but the land just seemed to swallow them up. We looked long and hard, scanning the horizons, but there was nothing to be seen.

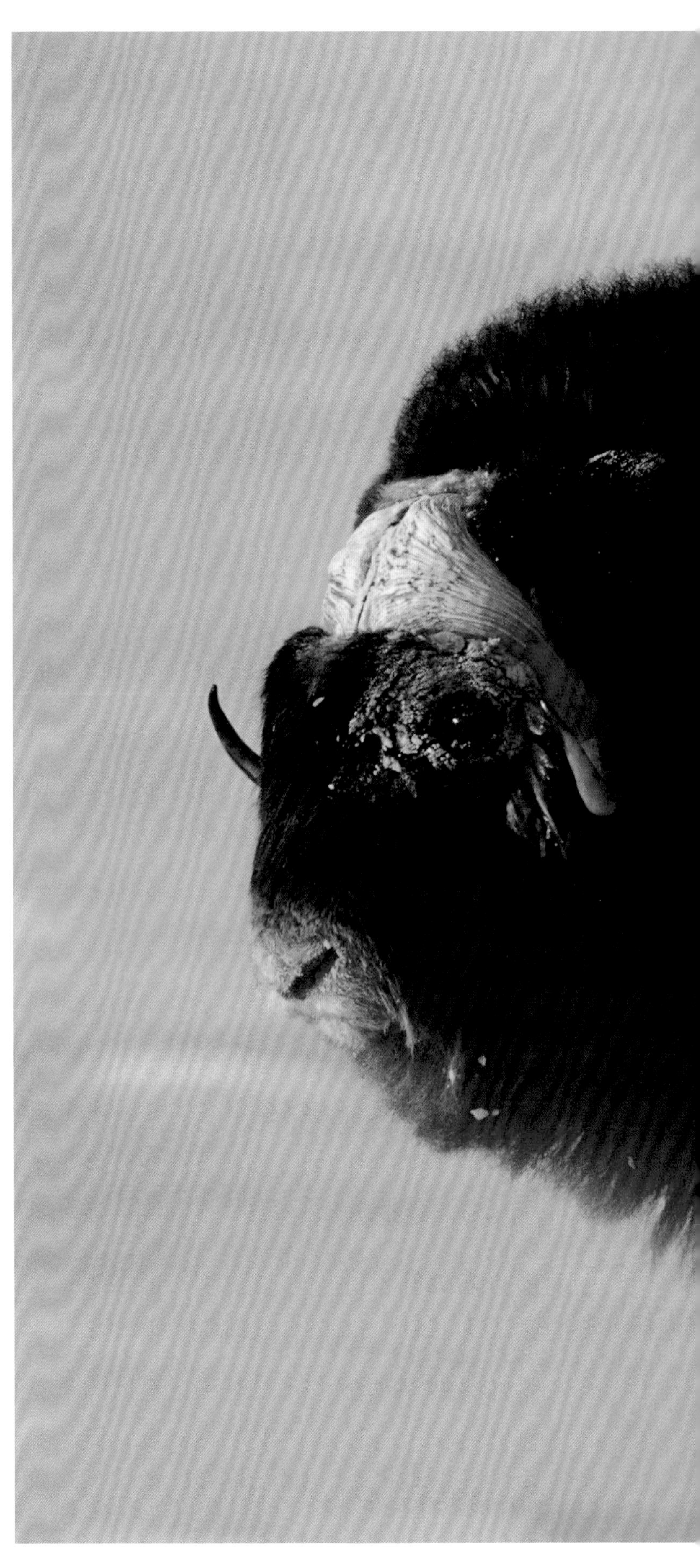

Aurora borealis (above): the northern lights are a common sight in winter skies. They are formed by solar winds interacting with the Earth's magnetic field.

Musk oxen (right): one of the hardiest animals in the Arctic, these bulky herbivores do not migrate south in the winter. To keep warm they have an underlayer of thick, fleece-like wool covered by an outer layer of long guard hairs.

The extent of the sea ice

The difference between the Arctic seasons is startling. The seas surrounding Svalbard freeze over in the winter, merging the islands into the permanent pack ice of the polar north. In summer this ice retreats, opening up a rich coastline of bays and fjords.

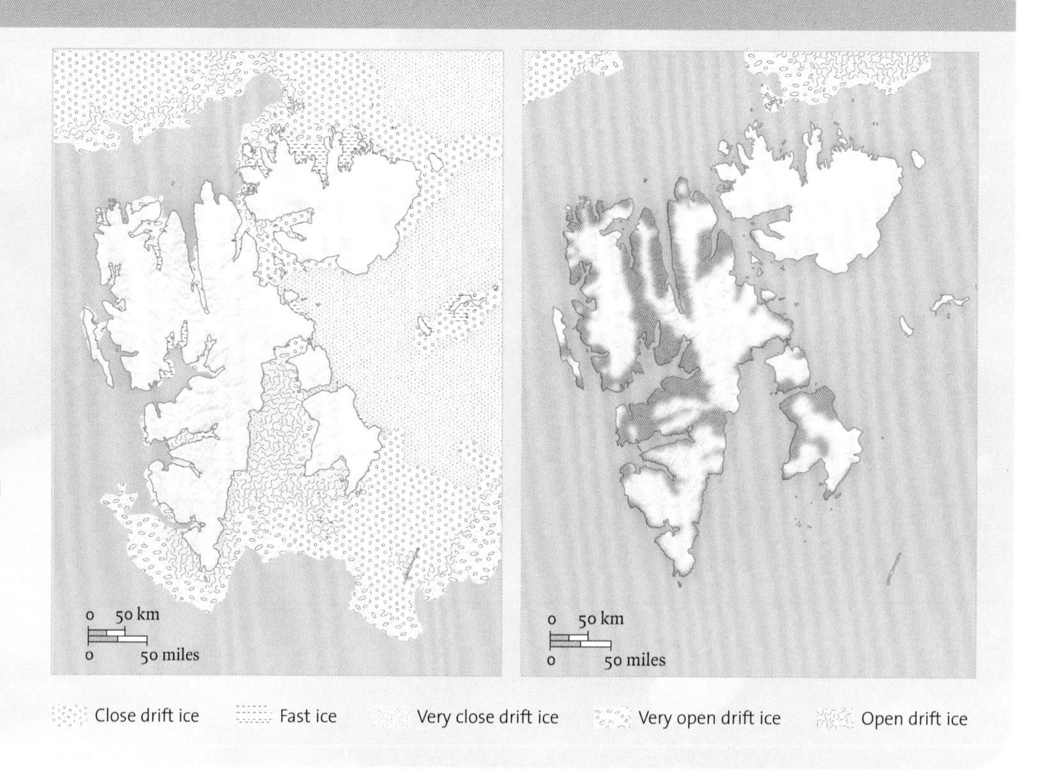

Winter (right): in the winter, fast ice forms in the sheltered bays and along the coasts. Drift ice is pushed from the north extending around the island's east coast. The drift ice can vary in density according to currents, winds and sea temperatures.

Summer (far right): across the short summer, the ice retreats. The bays and fjords are released from the grip of the fast ice and the drift ice melts away leaving open ocean.

The melt begins

For a few days before the first return of the sun, the Arctic sky is painted many shades of purple and blue by an ever-lengthening period of twilight. Then finally the sun makes its first appearance over the horizon. This first sunrise after the months of darkness remains one of the most beautiful sights I have seen in the Arctic; its light and warming rays raise the spirits. In Longyearbyen the sun rises and sets after only 1½ hours on the first day, but thereafter the length of daylight increases by roughly 40 minutes a day. The distance between the positions of sunrise and sunset on the horizon becomes further each day until the end of April, when for the first time the sun doesn't set at all. For human visitors from southern latitudes the permanent daylight can be difficult to deal with. On my first visit I found it almost impossible to sleep; without nightfall to define periods of necessary rest it became easy to keep working and lose track of what time of day it was. As a consequence I became permanently exhausted and took a long time to readjust to normal activity patterns when I returned south. It's a strange truth that the Arctic receives the same amount of sunshine each year as the tropics, the difference being that in the Arctic it all comes at once.

This increase in sunlight triggers the beginning of the great melt. In spring when the sun is still weak, the process is slow and there is often a period of melting and refreezing as the temperatures fluctuate.

Melting snow on the land often precedes the melting of the sea ice and for a short time in spring the land is crisscrossed by freshwater streams, fed by the melting snow. These streams flow into the ocean and dilute the salt water, speeding up the melt along a thin band of sea ice adjacent to the land. With the start of summer the Arctic becomes a place of strange contrast, the land covered with the fresh green of plants and flowers whilst the sea is still a frozen white desert.

In order for the melting of the sea ice to begin in earnest, temperatures need to be consistently above 0°C (32°F). As spring turns into summer, the sun grows higher in the sky and its rays hit the ice more directly. Half the incoming solar radiation is reflected off the ice, producing a shimmering landscape of heat hazes. Snowfall on the ice takes longer to melt, as the snow may reflect up to 90 per cent of the sun's power. As it does melt, it forms a maze of turquoise pools across the ice's surface. These meltwater pools get broader and deeper as the melt speeds up because water is less reflective than snow and ice, and it absorbs more heat.

The ice splits, cracks, and is pushed and pulled by the wind and current. The sun is not strong enough to melt it all. The thickest of the winter ice will survive, albeit reduced to small fragments and blocks. At the height of summer this creates an ever-shifting patchwork of drifting ice and open ocean.

The rugged landscape of Svalbard (next page): as temperatures rise, the last remnants of broken sea ice begin to drift offshore.

The melt begins (right): the process of the sea ice melt is not straightforward. With the sun's return temperatures generally increase, causing widespread melting, but cold snaps do occur and areas of delicately patterned re-freeze ice are formed.

The invasion of life

The summer seas of the Arctic are rich in life. The melting ice leaves behind it a thin, nutrient-rich layer of fresh water on the ocean's surface. Microscopic organisms feed on the nutrients and become the building blocks for a complex food chain. When sunlight penetrates the water, conditions are perfect for phytoplankton to bloom. These green algal carpets coat the underside of the cracking sea ice, providing food and habitat for shrimp, which in turn support huge numbers of fish such as Arctic cod. The fish are the reason millions of seabirds flood north to breed in the summer. Little auks, guillemots, puffins, kittiwakes, and fulmars congregate to nest on the steep vertical walls of coastal cliffs. With permanent sunshine and an excellent food supply, they have a great opportunity to rear their chicks.

Brunnich's guillemots (left) rest on the ice during a break from diving to catch fish. This species is amongst the most numerous and widely distributed of the Arctic's seabirds, with millions of birds in the north Atlantic and north Pacific polar seas.

Ocean currents

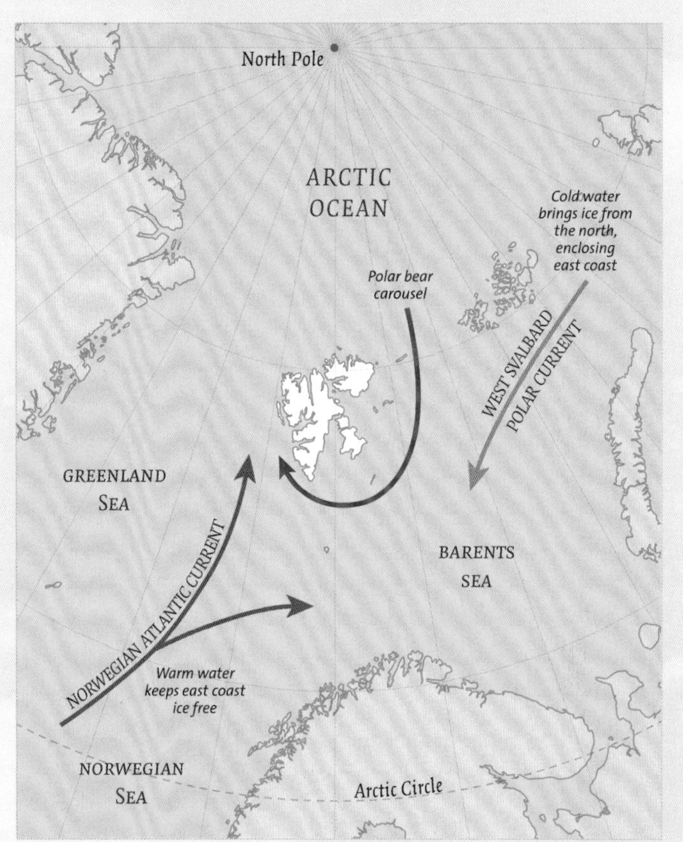

The growth and melt of the sea ice is greatly influenced by ocean currents. The east coast experiences high levels of ice due to cold water currents from the north, whilst the west coast, warmed by currents from the south, often remains ice free. In years of high ice levels, these revolving counter-currents draw ice south along the east coast and push it around the cape up the west coast. Polar bears, at the mercy of the ice movements, are drawn around the islands on this "polar bear carousel".

The large numbers of fish also support thousands of belugas, who swim over 1500km (1000 miles) from southern Greenland to Canada's Lancaster Sound to feed and produce their young. They travel through the fragments of ice, navigating their way into shallow bays where they can give birth in safety. The shallow water protects them from killer whales, but also allows them to rub their moulting skin on the rocks. In Cunningham Inlet on Canada's Somerset Island, as many as 3000 belugas gather in dense groups (see page 176–7). The Inuit say you can walk from one side of the bay to the other by hopping from the back of one whale to the next. It is one of the most unusual and impressive spectacles anywhere in the natural world.

On the land, too, there is a summer invasion. The partial melting of the permafrost awakens hordes of mosquitoes from the soil. Many people associate swarms of blood-hungry biting insects with the tropics, but I have seldom encountered a more persistent torment than the clouds of mosquitos that pestered our film crew on the summer tundra of Baffin Island. At times it seemed the only escape was inside a sleeping bag! This explosion of insect life supports a rich cast of migratory birds – as many as 70 species, including phalaropes, plovers, buntings, and larks, travel north to take advantage of the easy pickings.

In places the annual snowmelt leads to a fresh growth of grasses, which support huge colonies of grazing birds. One such gathering on Baffin Island is home to a staggering 1.5 million lesser snow geese, which have migrated all the way from New Mexico for the short summer. They moult in August and the local residents, who visit the colony to gather their eggs, describe a great open plain where the air is full of white feathers that fall like snow.

Belugas (above): in spring these highly social whales migrate north into cold Arctic waters. They are insulated from the cold by a 10cm (4in) thick layer of blubber. Up to 50 per cent of a beluga's body weight can be made up of fat.

Whale migration
Avoiding the ice

There are three species of whales that survive in the freezing Arctic conditions in order to make the ocean at the top of our planet their home. These are the ghostly white beluga (*Delphinapterus leucas*), the extraordinary tusked narwhal (*Monodon monoceros*), and the large black bowhead whale (*Balaena mystericetus*).

During the winter months the sea is frozen over and even Arctic whales cannot survive here as they are unable to reach the surface to breathe. But as the great melt begins, the sea ice starts to retreat allowing the whales to access the Arctic seas and make the most of the abundant opportunities.

Journey begins
In early spring, Arctic whales leave their wintering grounds south of the sea ice and gather at the ice edge to watch and listen for signs of the start of the melt. Shore-fast ice – forced by strong winds, upwellings, and tidal currents – fractures, ruptures, and tears open to reveal long stretches of open water known as leads. Some can be so broad that dozens of animals can swim abreast, whilst others are narrow to the extent that the whales are forced to swim head to tail on their journey north.

Love in the air
In April belugas and narwhals collect in their thousands to mate in the offshore ice of Baffin Bay and Davis Strait, waiting for the ice of Lancaster Sound and the surrounding inlets to give way to the prevailing conditions. Despite the sub-zero temperatures the whales are full of the vigour of spring; this is the time when males compete for females and mating takes place.

Bulldozers of the north
In late April, Western Arctic bowheads lead the way, blazing their own trail. By distinguishing between different densities of ice and using their huge heads as powerful battering-rams, bowheads carve themselves access to the air they need. They often travel in small groups of up to 15 whales that spread out over 10–15sq km (4–8 square miles) and remain in contact by calling to one another. The belugas follow the bowheads north, sometimes directly on their tails, making the most of the routes the bowheads have found.

Imprisoned by ice
Occasionally, shifting ice or narrowing leads trap migrating whales, forcing them to remain at small openings for perhaps weeks at a time. The whales' repeated surfacing helps to keep the holes ice-free. Inuit hunters call these whale traps *Savssat* – for the local people they can be an important source of food at this time of year. *Savssat* also provide patrolling polar bears with opportunistic meals, though such large, water-born prey is not easy to catch!

Flood gates open
In June and July, when channels through the ice are finally open and the high (Lancaster Sound, Baffin Bay), eastern (Cumberland Sound and south-eastern Baffin, Hudson Bay, James Bay and Ungava Bay), and western (Beaufort Sea) Arctic have finally opened up, whales flood in, drawn by an abundance of food. From these places they disperse into adjacent fiords, inlets, and bays, to nurse their young and to feed and socialize.

Southerly retreat
Winter signals its approach with shorter, cooler days, and in response the Arctic whales prepare for their autumn migration before the sea ice blocks their path. In the high Arctic the whales begin to leave as early as mid-August, whereas those summering further south in Hudson Bay don't need to leave until early September.

Bowhead (above)
At 15–18m (50–60ft) long, the bowhead is the largest of the Arctic whales and yet it only feeds on plankton. Bowheads are named on account of their bow-shaped mouths and massive heads – which make up 40 per cent of their body length – and it is this unique physiology that allows bowheads to survive in the Arctic longer than either belugas or narwhals. They can break through ice 22cm (9in) thick using a combination of a powerful forward momentum and their immense 60–80 tonnes of body weight, and the impact of this collision is cushioned by a thick pad of fibrous connective tissue on the top of their heads. Between breaths they can travel for long distances under the ice at 3–6km/h (2–4 mph). These true Arctic explorers can live to well over 100 years – in some cases up to 200.

Beluga (bottom right)
Belugas are very social whales that earned their nickname "sea canaries" from early whalers because of their constant chattering. At about 4.5m (15ft) in length, the beluga is a relatively small, toothed whale which, unlike most other Cetaceans, has seven neck vertebrae that are not fused, giving belugas a flexible, well-defined neck. Belugas seldom breach but rather bounce vertically, with about a third of their body length leaving the surface.

In summer thousands of belugas collect in estuaries on the western side of Hudson Bay and in Cunningham Inlet in the high Arctic. They come to these warm, brackish waters to feed, moult their skin on the rocks, and nurture their calves. It is also thought that the belugas congregate in shallow water because of the relative protection from larger killer whales, which cannot follow them into the shallows.

Beluga skin is a very dynamic organ, which scientists have only recently shown is shed every year. It is very thick – about 10 times thicker than that of dolphins and 100 times thicker than that of terrestrial mammals – and is used for insulation, storage of high quantities of vitamin C, and, possibly, protection from the abrasion caused by contact with ice.

Narwhal (Tusk) (above)

Like the beluga, the narwhal is a small whale, but it has one remarkable feature. Male narwhals – and the occasional female – possess impressively long, unicorn-style tusks in the centre of their foreheads that make them look something like a mythological creature. In males the tusk can weigh 9kg (20lb), and be more than 2m (6ft 6in) long and 25 cm (10in) in girth.

The function of this extraordinary tusk has been debated over the years, with many scientists believing it is simply a weapon for jousting males. But recently, a new study made a startling discovery: the tusk is a sensory organ of exceptional size and sensitivity that can detect subtle changes in pressure, temperature, particle gradient, and probably a lot more. Ten million tiny nerve connections from the central nerve of the tusk to its outer surface enable this sensitivity. It is now believed that the tusk's main purpose may be to locate prey, and with no easy way of cleaning such a tool, it may be that the "jousting", where males gently brush tusks, is simply a means of removing grime. The whales aren't fighting, they're cleaning their teeth!

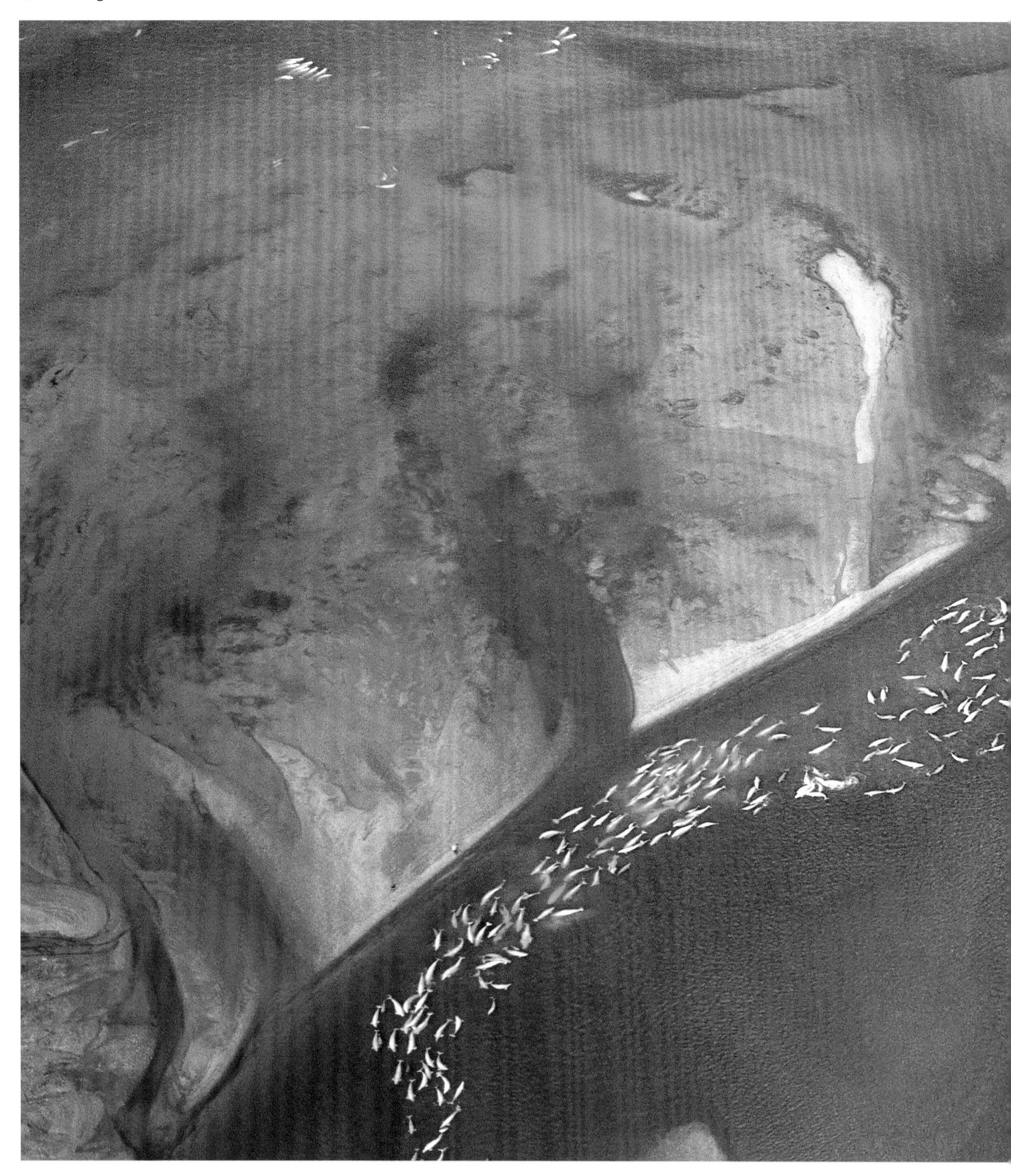

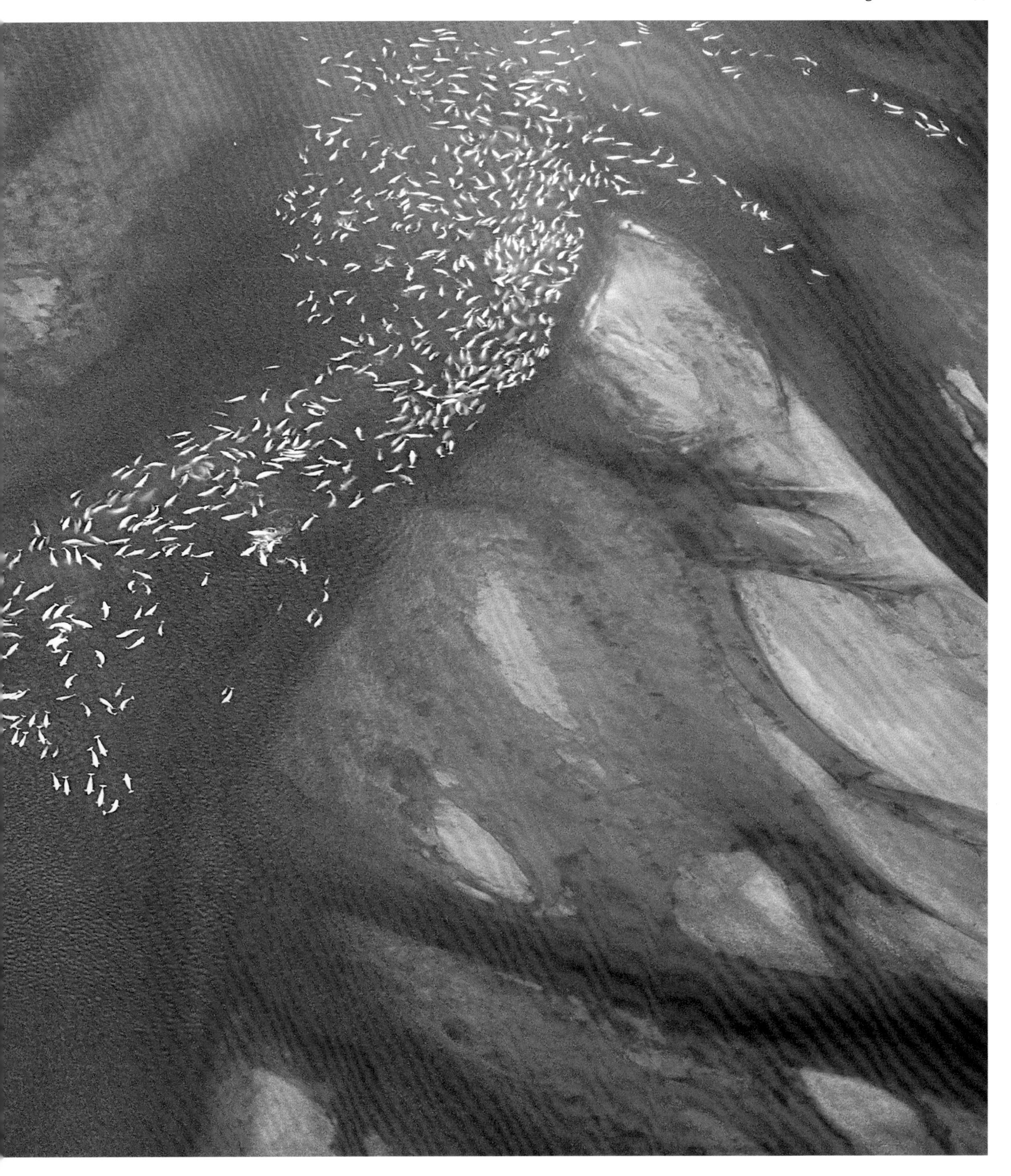

Boom and bust cycles

Lemming (above): these little rodents can produce a litter of six young every month throughout the summer.

Arctic hare (above): individuals from the most northerly islands remain white all year round.

The beginning of the melt on land brings about a sudden abundance of light, water, and fresh organic nutrients that leads to a concentrated period of bloom and boom. Within days grasses, lichens, birches, and flowering plants begin to grow.

Lemmings and Arctic hares emerge from hibernation in time to gorge themselves on the fresh pastures, providing many predators with an ample food supply. Populations of lemmings rise and fall in a roughly four-year cycle, though the exact timing and the degree of fluctuation depend on the harshness of the preceding winter and the timing of the melt. Around Churchill, Manitoba, one lemming species varied in density from 40 adults per hectare (15 per acre) in peak populations to approximately one individual per 15 hectares (5.8 acres) in a low year – a more than 600-fold decrease!

Lemmings are at the bottom of several Arctic food chains and during boom years many predators benefit from the abundance of bite-sized rodents wandering the tundra. Stoats, wolverines, wolves, foxes, gulls, owls, and even reindeer have been recorded eating lemmings. The fortunes of all these predators are tied into the four-year cycle. When lemming populations are high, snowy owls, Arctic foxes, skuas, stoats, and other predators – some of which have migrated to the lemming-flush areas – produce more offspring. But after a boom must follow a bust: predation, disease, and overgrazing cause the lemming populations to crash. In the following years predators either produce fewer or even no young, or they emigrate to better lemming areas, allowing the lemming population in the original area to recover.

With the onset of the melt, disappearing snow saturates the soil and vast expanses of the tundra become covered with a network of freshwater pools. These pools play host to one of the Arctic's most amazing population explosions. Female mosquitoes prefer to lay their eggs in the soil on south-facing banks of tundra ponds. These sites receive more sunlight and the winter snow melts here first. The eggs are laid just above the summer water level. The melting snow submerges the eggs buried in the soil and once underwater they hatch out. The young larvae take just short of a month to mature and soon the tundra is buzzing with clouds of blood-sucking insects. This seasonal explosion in insect numbers draws in a multitude of small bird migrants.

Snow geese (above): for the many Inuit and Inuvialuit residents of the Arctic, the arrival and departure of these huge flocks mark the beginning and end of the great melt.

The Great Melt, therefore, is not only one of the planet's largest physical seasonal changes; it is the destination for one of the most dramatic movements of animal life, large enough to be virtually unquantifiable.

But whilst the melt provides great opportunities for some animals, for polar bears it is a more difficult time. With the sea ice they rely on to hunt for seals now melting, they must follow the last fragments of shifting ice if they are to hunt. Catching seals is much more difficult in the summer. The bears must hunt well before the ice has broken up, in order to put on enough calories to survive the summer famine. For polar bears it really is "survival of the fattest".

In order to film the great melt the BBC would spend two years in the Arctic wilderness to capture the magic of one of Nature's Great Events.

The main location for filming was to be the Norwegian archipelago of Svalbard. Cameraman Martyn Colbeck and I flew north to begin a quest to film the Arctic's most famous resident – the polar bear. It was spring and, although the sun had returned, the land was covered with snow and the sea was still frozen. The melt had yet to begin and for the bears this was a crucial time. With the island surrounded in sea ice, they would be busy hunting for seals. Our goal was to reveal something of their behaviour in this winter wonderland.

Snowy owl (opposite top): the birds' breeding success is linked to the population cycle of the lemming. The owl lays between three and 14 eggs and survival depends on how many lemmings the birds can catch. In a low lemming year only a single chick may make it.

Cunningham Inlet (previous page): hundreds of belugas rest in the shallow waters off Somerset Island in Canada. The whales come here to moult in the warm water and rub their skin on the smooth rocks of the river bed.

Out in the cold
Justin Anderson

We had only just steered our snowmobiles past the outer limits of Longyearbyen town when the weather closed in. All morning a cold blue sky had helped our final efforts to pack six sledges with a mountain of equipment; now low white cloud rolled down the valley, driven on a piercingly chilly wind that blew straight into our faces. At this time of year the weather could change very quickly, sunshine and blue sky turning to storm and whiteout in an instant.

Leading our group into the whiteout was Jason Roberts, an experienced Arctic guide and location manager for our filming. Jason's snowmobile hauled two huge sledges piled with filming kit, food, and cold-weather survival gear. I went second, glad to have only one sledge to pull as I was still getting used to the steering, throttle, and brake, and didn't want to make any emergency stops and jack-knife due to the weight behind me. Next came Martyn, on his first visit to the High Arctic and excited to be heading away from town and towards the chance to see his first polar bear. Bringing up the rear was Steinar Aksnes, an experienced polar traveller and our second guide.

As the blizzard closed in around us, Martyn felt a strange stinging sensation on his neck and brought his snowmobile to a halt to investigate. He pulled off the many layers of hoods and ski masks, scarves, and balaclavas to reveal a block of ice sitting under his chin. There had been a gap in his clothing and the icy winds had done their worst, freezing a patch of skin until it turned white. It was a good lesson that even within sight of town we had to be vigilant about covering up and checking on each other. The ambient temperature was –20°C (–4°F) and the combination of wind chill and the additional airflow from riding the snowmobiles meant any uncovered skin could freeze in a minute.

In order to survive and work in these conditions, each of us wore huge piles of specialist clothing: a base layer of thin woollen thermals, woollen socks, thin fleeces, an all-in-one snowmobile suit, three pairs of gloves,

En route over the Paulabreen glacier (left): we used these high glaciers as "roads" to cross the island on our snowmobiles.

and a variety of hats, neck guards, balaclavas, snow goggles, and specialist boots with heat-retaining reflective layers. All in all we looked better suited for outer space than a film shoot.

Before us a gently rolling expanse of white stretched to the horizon, the glacial highway that would lead us to our destination. All along its edges, serrated lines of snow-covered mountains completed the fantasy landscape. The sun did little to warm the freezing air, which I could feel pinching at my cheeks and lips as we hurtled across the new powder snow.

We worked our way down through some tricky glacial moraines to our destination: the east coast of the main island of Spitsbergen and the heavily glaciated bay of Kvalvagen, which would be our home for the next four weeks. For as far as I could see there were peaks and bays, each with a huge glacier emptying into the sea. At this time of year – the early spring – the polar current pushes sea ice southward from the far north, closing in bays and the fronts of glaciers, locking the coastline in an icy grip. Once the ice gets far enough south it will round the southern cape of Spitsbergen and the Gulf Stream will propel it north again, so that as winter draws on, the whole island may be enveloped in ice. Polar bears ride this sea ice "carousel" and the conditions are ideal for them to hunt for seals. At this time of year Svalbard plays host to 3000 "isbjorn", or ice bears, as the locals call them.

Judging by the huge number of footprints around our small cabin, most of those 3000 seemed to live close by! But luckily neither bears nor bad weather interrupted us as we began the frantic preparations to get ourselves and our gear into the hut before the sun sank low and the temperatures dropped. Jason showed us how to set up a bear fence: a trip wire attached to a series of firework explosive charges that would hopefully scare off any bear attracted by the smells of our cooking.

Once we were all inside, the hut was very cosy, if a little cramped. With four of us squashed into 10sq m (100sq ft), we would have to get used to living in close proximity.

Our first foray

In the morning we set out to look for bears to film. Jason led us down into Kvalvagen Bay, where three huge glaciers empty into the sea. The sea ice in these glacial bays is known as fast ice, and it is the thickest and most stable sea ice around, thanks to a combination of cold winds off the glacier, heavy snowfall, and sheltered conditions. The fast ice is the favourite place for ringed seals to build their birthing dens.

The seals pup from the end of March to the beginning of April and dig upwards from under the sea ice, building a chamber under the snow where

Following paw prints (right): a good way to find polar bears on the sea ice. Here two tracks become one, as a mother bear gives her tired cub a ride on her back.

Getting about (above): Snowmobiles are the best way to travel across the vast expanses of frozen land and sea, looking for polar bears.

Essential protection from frostbite (above): cameraman Martyn Colbeck, wrapped in many layers of warm clothing.

Home from home (above): the small wooden hut at Kvalvagen where we lived for a month during the winter filming.

Ringed seal and pup (above): the main prey species for the polar bear. Seals give birth to their white-coated pups in lairs beneath the ice in the spring, a key time in the bears' fortunes. The pups can be up to 75 per cent fat, a rich and abundant energy source for polar bears.

they can give birth. Not only are the pups kept sheltered, but they are also out of sight of predators such as Arctic foxes and polar bears. Should a bear smell a seal's lair though the ice it must break through the snow on the surface, giving the seals precious time to escape through a hole in the ice to the ocean below. This is one of a number of reasons why only one bear hunt in 20 is successful!

We snowmobiled long distances over the fast ice, following the glacier, towering blue walls of ice looming over us. Jason spotted something through his binoculars and we headed onwards. Around the corner we found the large tracks of a mother bear and the smaller, more rounded tracks of her cub. We followed them until the cub's tracks disappeared. Jason told us that the mother bear must have given a lift to her tired youngster, letting him climb up on her back and rest as she walked on (see page 182–3). We tracked the footprints until they headed out onto thin ice where we couldn't follow.

Bears – up close and personal

Lying in my bunk I read the chapter on polar bears of Svalbard in my travel guide. Unfortunately it seemed preoccupied with horrific bear attacks and it left me a little nervous about our filming. One grisly tale told of a group of Austrian climbers who in 1977 had a bear come into their camp in Magdalenefjorden. The bear attacked a man coming out of his tent, dragged him out onto the sea ice and ate him in full view of his horrified companions. Another story concerned two students who decided to climb a small peak above Longyearbyen. They saw a distant animal with white fur and approached, thinking it to be a reindeer. By the time they realized it was a bear it was too late. One of them ran and the bear chased after her. Her friend slid to safety down a rocky ravine and raised the alarm, but the bear attacked the rescuers rather than give up his prize; in the end the bear had to be shot, but the girl was already dead.

Jason told us that all bears are unique and that one bear's reaction to an encounter with humans may be vastly different from the next's. One might turn and head off, another might be curious and approach to a comfortable distance, still another might keep coming, and we should always be prepared to make our exit. If we came across a curious, fearless bear, most likely it would be an adolescent, struggling to hunt successfully

in its first winter and desperate for any food. In most of these scenarios the starting of the snowmobile engine would be enough to discourage it, but Jason carried flares and pepper spray and, as a very last resort, there was a rifle strapped to his snowmobile. Jason reiterated that at all costs we would avoid disturbing the bears or causing them stress, only filming those individuals who were relaxed in our presence.

So that Martyn could set up for filming quickly when we encountered a bear, Jason's snowmobile pulled a sledge onto which we had strapped our tripod and camera box. It also meant that, should a bear get too close and keep coming, Martyn could hop on the sledge and Jason could drive him away to safety.

There were no bears to be seen around Kvalvagen Bay, so we headed north, cutting inland to circumvent the glacier snouts and precipitous cliffs before turning back down to the coast again. We emerged onto the fast ice of Inglefieldbukta Bay and soon spotted an old male bear, his hip and shoulder bone poking through his fur, his neck long and shaggy. Still as a statue on

Seals are hard to catch (above): despite the bear's skills as a hunter, only one in 20 attempts ends in success. The crew watched this bear catch a small ringed seal in the summer pack ice. Here it is pulling off strips of energy-rich skin and blubber.

Life under the ice

The Arctic Ocean is a unique place. On the shoreline shifting ice scrapes away any limpets and mussels, and further out, icebergs gouge rifts in the seabed. During the coldest winter months there is little or no sunshine penetrating the water and in summer the sun may shine 24 hours a day. Animals that can survive these conditions often have very special adaptations.

When sea ice forms, small spaces between the ice crystals remain and are filled with brine and organic matter, which are slowly released once the ice melts. Cracks and edges become epicentres for a rich diversity of sea life; sunlight penetrating the water accelerates further melt, and an early growth and multiplication of microscopic algae and fungi are triggered. The algae then fix solar energy, in a similar way to plants, forming the basis of the marine food chain.

During summer enormous swarms of sea snails and jellyfish travel to the Arctic on the North Atlantic Drift. One unusual sea snail is the winged *Limacin helicina*. This 15mm (⅝in) long "flying" snail develops as a male, but becomes a female later in its life. The largest known jellyfish is the 2.1m (7ft) long Arctic lion's mane jellyfish. Covered with millions of stinging cells, it is a very efficient predator.

The summer seas of the Arctic are also rich with shoals of fish. Deepwater Arctic cod is one of the most important links in the Arctic food chain – it feeds on the zooplankton that feed on algae – providing the staple diet for many larger fish, seals, and whales. Almost uniquely, the Arctic cod has adapted to survive here by developing "antifreeze" glycoproteins in its blood that allow it to freeze itself during the winter and thaw out in summer.

Because of the ice, the Arctic Ocean remains the least explored region of ocean on the planet. Only now are scientists beginning to understand the potential diversity of this "blank on the map".

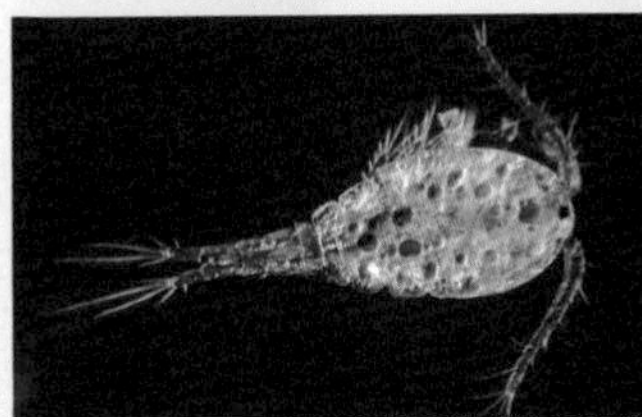

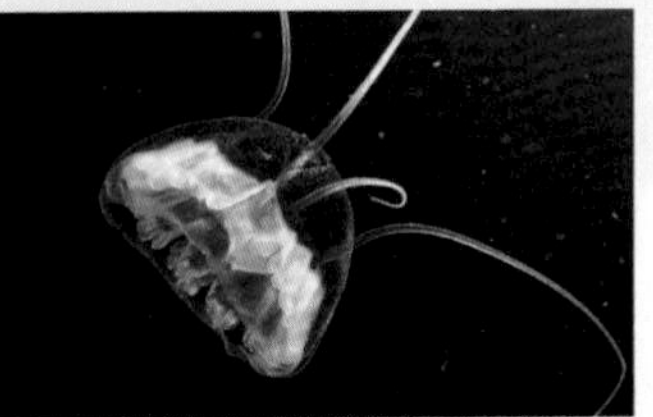

Arctic cod (top) spend their entire life in temperatures close to and below freezing.

Copepods (above left): the most numerous zooplankton in the Arctic waters, these crustaceans overwinter in deep waters at 300m (1000ft), but as soon as the ice breaks and the sun penetrates the surface they head upwards to feed on the algal blooms.

Jellyfish swarm in the summer waters (above right): millions of tiny 4cm (1½in) comb jellies and sea gooseberries drift on the currents.

the ice, he stood over a seal's breathing hole, waiting for the seal to surface. Hunting bears can wait patiently like this for hours, but this one looked uncomfortable: every now and then he would gingerly adjust his position or lie down to stretch out his stiff back. Martyn got some lovely footage before the bear finally gave up on the hole and headed off out to sea. We followed at a respectful distance, but Jason slowed us to a halt and warned us that we were now on "overwater" and should move quickly to solid ice.

Overwater occurs where thick fresh snow falls on thin sea ice. The resulting thermal blanket turns the ice below to slush. The snow and slush can often support quite a bit of weight, but where the ice is thin, a sledge runner or a boot can go through and get a soaking in the water beneath. Driving on overwater was always an unnerving experience. As we followed Jason to firmer ice, our route took us close to the bear, who barely raised his head at our passing. I was too absorbed in watching him to notice that Jason, with Martyn perched on the filming sledge, had nosedived his snowmobile through the overwater. Unable to react quickly enough, I ploughed into them, locking my runners into those of the filming sledge.

We sat there trapped by the ice and locked together. I looked back to the old bear who, 40m (130ft) away, had suddenly been presented with the possibility of an easy meal. A nervous few moments passed until, no doubt unimpressed with our performance, he nonchalantly turned away and headed out to the ice edge without looking back. Relieved, we set about hauling the snowmobile and sledge apart.

Frustrated by the weather

The following morning, we woke late to a blizzard that continued unabated for two days. Occasional trips outside to pee or dig out the snowmobiles meant we returned to the hut covered in snow, beards and eyelashes frozen. On the evening of the third day the blizzard finally eased, but clear skies overnight sent the temperature plummeting to –25°C (–13°F), the coldest yet. The sea ice had frozen further out from shore and seemed to extend all the way to the horizon. Driving on this new surface was a strange experience – only a day or two, before it had all been open ocean. Steinar was very careful to measure the thickness of the new ice every 20m (70ft) or so with an ice screw. Anything less than 25cm (10in) was deemed too thin for safe travel. As we crossed one stretch of flat grey ice, the pressure of the snowmobiles forced it to ripple, a shallow wave stretching out before us. At this time of year, conditions alter rapidly. The days grow slowly warmer but there are cold snaps too, meaning the ice is very changeable: one minute open water, the next solid sea ice. This unpredictability made it hard to plan

The search for polar bears (left): looking for filming opportunities took the crew through an incredible landscape of mountains and glaciers.

our routes; many times we would find a bear only for it to lead us towards ice that was too thin to bear our weight.

A bear called Scar Nose

Around the corner from the hut, heading north along the coast, we discovered a huge sea cliff with a broad rocky ledge part way up. We could access the ledge by way of a steep ramp of snow, which could be scaled only by taking a long run up with the snowmobile and flooring the throttle to get the necessary momentum.

The view from the ledge was incredible. As we stared over the frozen sea, Martyn dreamed out loud, "Wouldn't it be great to film a bear from up here?" It was a great spot for a lapsed time position, and we planned to return in the summer and film from the same viewpoint, giving us a direct comparison between the seasons. As we stood on the ledge, a few graceful fulmars swept out of the air, gliding in low to look at us, curious as to who the strange visitors to their cliff might be. These pioneering birds were the first arrivals here. Having spent the winter feeding in the ocean further south, they now sought out the cliffs of Svalbard to mate and rear their chicks; they would rather put up with the cold than miss out on the best ledges.

We had returned to the hut and were setting up another lapsed time position when Steinar alerted us to a large male bear walking south along the coast. We set off in pursuit and captured some wonderful images of him walking through a shimmering heat haze out on the sea ice. Then we set up the camera on the top of some moraine cliffs, hoping that he would come close to the shore if he came into the bay. He didn't disappoint, cutting in off the sea to walk just below us about 50m (165ft) away. He was huge, about 225kg (500lb), with an enormous square head. His neck was covered in scars and scratches, no doubt from fights with other males. The most prominent scar was across the bridge of his nose, giving him the battered look of a heavyweight boxer.

Scar Nose didn't pay us the slightest attention; he seemed completely focused on sniffing the floor, stopping to hoover up information, his

Scar Nose (above): this large male bear proved to be a wonderful character, allowing the crew to film him up close. The prominent scar on the bridge of his nose is likely to be from fighting with other males over females.

The changing seasons: in the course of filming, the crew was privileged to see the huge differences in the landscape. Where once there had been ice to drive on **(above left)** there were now crashing waves and open sea **(above right)**.

Polar bear
Ursus maritimus

A polar bear's sense of smell is one of its most formidable features: at least 20 times stronger than a human's, a polar bear can smell a seal's lair from over 1km (½ mile) away. Cubs are born in dens under the snow, and break free of the den in the spring. Over the three years a cub spends with its mother it learns to hunt, feed, and interact with other polar bears by watching its mother intently, often mimicking her behaviours and movements exactly.

The secret to a polar bear's survival in the Arctic is insulation and maintenance of a constant body temperature. Polar bears have a number of adaptations that enable this, including a layer of blubber up to 15cm (6in) thick, dense under-fur at least 5cm (2in) long, and an outer layer of hollow guard hairs that grow to 15cm (6in) and act like a down sleeping bag, trapping air for further insulation. The fur is surprisingly water-resistant, so that when a bear emerges from swimming it simply shakes or rolls in snow to dry off.

KEY FACTS

Location	Arctic, northern Canada
Habitat	pack ice, shore line, open water.
Lifespan	up to 25 years
Size	2–3.5m (7–11ft)
Weight	300–800kg (660–1760lb)
Food	seals, young walruses, whales, reindeer.

black nose pressed onto the ice. Occasionally he found a smell so intoxicating that he would roll in it, rubbing his sides and back into the ice. Jason had seen this kind of behaviour before and told us that it looked as if a female on heat had come this way, following the coast. Scar Nose was in hot pursuit.

It was clear that this was the peak of the bears' mating season. A few days earlier we had been privileged to witness the courtship ritual of two bears we encountered on the ice. The female rolled in the snow, never letting the male get too close. He would catch up and roll on the same patch, his black feet kicking in the air, eyes closed tight. She gradually allowed him to come closer, running a little ahead, controlling the distance, until finally they gently touched noses and mated. After mating the male bear will follow the female for several days, seeing off any potential competitors. Sometimes well-matched males fight viciously for access to females and this evolutionary pressure has led to the large difference in sizes between the sexes: male polar bears weigh 350–800kg (770–1750lb), females only 150–300kg (330–660lb).

We soon realized that Scar Nose was heading north and there might be a chance to get ahead of him and up onto the fulmar ledge to film him as he walked out onto the sea ice. We skirted along the coast and cut inland, but the bear's seemingly plodding gait was deceptively good at eating up the ground: even on snowmobiles we struggled to keep up.

We roared up the snow slope and onto the ledge, then set up the camera and waited, eyes trained out to sea. I wondered what would happen if the bear didn't stick to the coast but followed us up onto the ledge: we

would be cornered! Quickly banishing the thought, I went back to looking for him through binoculars. Right on cue old Scar Nose came around the corner, no more than 60m (200ft) or so below us. It was an excellent viewpoint, allowing us to film without him even knowing we were there. He reached an area of refrozen ice and stepped gingerly out onto it. Still sniffing at the floor he walked on, following the female's scent, but the ice began to wobble under his footsteps. Sensing this, he spread his weight, stretching out his front and back legs and sliding with his paws over the thinning ice. Too late, the surface crumbled and the bear disappeared into the water with a gentle splosh. Up on the ledge we all gasped, but we needn't have worried – Scar Nose took it in his stride, shaking lumps of ice off his head, and ducking below the water to see if he could swim to safer ground. After a few seconds he resurfaced and began to claw his way across the brittle new ice, which splintered and crumbled under his paws as he pushed down on it. Finally he pulled himself out and shook the water from his fur, then rolled on his back in the snow, using it like blotting paper to soak up the excess water. As if the whole thing had been a trifling inconvenience he padded off over the ice. At this point he got wind of us, turning and lifting his head to take in our scent, before heading out to the far reaches of the frozen horizon, drawn on by the scent of the female, which was in all likelihood more appealing to him than ours. I said a silent "thank you" to Scar Nose for being so tolerant and letting us share his day.

The great melt begins

Bad weather kept us stuck in the hut for another two long days. About mid-morning on the first day another curious bear came out of the blizzard and Martyn had to scare him away with a flare shot. The temperature outside climbed to –7°C (19°F). The winds rose and fell, changing direction constantly. By the time the bright skies returned, the ice in the bay had changed. A long line of thick slush had formed along the coast where the sea swell crushed the ice against the shore. Out to sea low clouds of mist hung over inky black water, now visible through giant cracks or "leads" in the sea ice. In the bay there were more patches of overwater and we drove at full pelt to avoid getting stuck. In places the ice was split by narrow cracks and we took it in turns to "jump" the gaps, thumbs pressed on the throttle.

A sound like a distant rifle crack had my head spinning to pinpoint the source. A small cloud of snow drifted from the front of the glacier, about 1km (½ mile) away. Several tonnes of ice had calved off the glacier wall onto the sea ice below. The echo took a while to die down, but was followed by the eerie sound of the ice creaking on the shore. Then came the strange

On thin ice (left): the recently frozen ice was not thick enough to support Scar Nose. Even though he spread his weight on his paws to cross a patch of new ice, he was still too heavy for it and he fell through.

sensation of rising and falling as the pressure wave from the impact travelled under our feet and bounced off the shoreline. The ice seemed very different and with the wind picking up Steinar was worried about us drifting out to sea on an ice floe: "It can change in an instant," he warned us. We headed quickly back to the hut, the clouds closing in behind us.

That night we woke several times with a start, convinced the bear bangers were firing, but it was only the thumping of the wooden beams in the hut, expanding as the temperatures climbed. Morning brought a raging blizzard and outside the temperature read only –3°C (27°F). Even with the low cloud we could see the dark stripe of open water in the bay. The sea ice had begun to open up. Cracks had widened and the tide and currents dragged the pieces of ice apart. Things were changing fast.

We had to content ourselves with waiting out the weather, while Steinar read to us from the local newspaper. The main story told of a trapper in a hut in the north of the island who had had a lucky escape. A polar bear had come through the window of the hut while he was sleeping and grabbed at the bottom of his sleeping bag. The man had defended himself by whacking the bear over the head with the blunt end of an axe head before it ran off with his bag. For once I was glad to be on the top bunk!

About 5.30 the following evening the weather cleared and we headed out to look for bears. The bay was now a patchwork of open water, flat ice, and chunks of glacier ice. There was a new sound of gentle surf rolling against the shore. Ice crystals floated on the air in the low sun and a cold fog skirted the distant peaks and shrouded the glacier front. There was a beautiful stillness to the world.

We picked up a male bear heading north along the coast. Could he be following the same female scent as Scar Nose after all these days? Again, we decided to get ahead of him and try to film him from the fulmar ledge. I had problems with my snowmobile: I hadn't given it enough throttle to make the slope and it became grounded in the snow. Finally reaching the top of the hill I was worried that my screw-up might have ruined a golden filming opportunity. So I was delighted to hear Martyn shout, "There he is."

Just a week on from when we had filmed Scar Nose walking across the frozen sea, we sat and watched another bear follow the same route, only this time the scene was very different. Where there had been sea ice there was now open water and an angular mix of cracked ice fragments and slush. Here and there huge blue icebergs were grounded on the low tide. Our bear was completely at home in this topsy-turvy environment and we

Naturally curious (above): polar bears will investigate any strange smells, including those from human dwellings.

watched him dive beneath the thin ice, his back bumping along its underside, white fur visible through the viscous layer, urgent snorts of breath passing through his nostrils as he swam. At one point he dived and disappeared. A nearby raft of floating seabirds panicked, taking to the wing, and he resurfaced, too late to catch one.

The first migrants arrive

We were back on the ledge at 3am. It was bitterly cold again but the sky was clear and there seemed a good chance that we could film a sunrise. Slowly the sun climbed over the horizon, lighting up a magical landscape. Clouds of golden fog drifted over the sea, vapour rising like mist from distant patches of open water. In the cold of the night, new ice had turned the surface of the sea to a thick soupy layer, the sunlight picking out the patterns of the ice crystals, as if the sea was made of liquid gold and mercury. Across this landscape paraded a most impressive sight: V-shaped flocks of guillemots, strung out in lines through the golden mist. Busy flocks of little auks wheeled overhead, their tiny wings working overtime to keep their stumpy bodies in the air. Gangs of black guillemots flashed their

orange feet in the sunlight, pecking at their neighbours for ownership of the best ledges. The whole cliff had come to life with seabirds from far and wide, here to make the most of the short summer in a place where rich seas and round-the-clock feeding were seabird heaven.

This amazing vantage point was to give us one last great show. A female bear and her newborn cub wandered out onto the thinnest ice. The cub was tiny and could only have been out of the birthing den for a few weeks at most. He ran around full of curiosity, skidding on his rump, and chasing and biting balls of snow. At one point he got the shock of his young life as his bottom popped through the thin ice, sending up a splash of water. He stared at the offending spot until a gentle huff from his mother called him onwards. It seemed unusual for a mother to be leading such a small cub onto the thin ice, but perhaps the scent of males in the area was making her keen to avoid the shoreline. Should a male catch up with her he would try to kill her cub to bring her back into mating condition. I couldn't help but wonder what the future had in store for this tiny bear and the others we had seen. The land and sea were changing fast and a new cast of

Warming up (above): little auks gather in huge numbers on the beaches of the Tusenoyane, or Thousand Islands.

characters were arriving. The great melt had begun in earnest and there would be lean times ahead for the bears.

With the ever-lengthening days and the increasingly early starts to film the morning sun, it's fair to say our circadian rhythm was somewhat confused. Before we went to bed at 8am after a full day's work I confessed to Martyn that I could no longer tell if it was today or tomorrow.

"It's neither, mate," he replied. "It's the day after yesterday."

Not sure whether to have a glass of wine or a bowl of muesli I gave up and went to bed.

Our final day began at 4am the following morning. It was a rude awakening. After a month of solitude in the remote wilderness, the figure of a huge man, dressed all in white, suddenly appeared in our small hut. I looked up at him with bleary eyes.

"Steinar," he boomed, "there is a polar bear in your garden."

Thinking it all a surreal dream, I closed my eyes and went back to sleep, only to discover later that our visitor had been 100 per cent real. Tommy was a trapper out of Longyearbyen and had gone for an early morning snowmobile ride to check out the Kvalvagen area. From a distance he had recognized Steinar's vehicle parked outside the hut and decided to pay us a visit. As he drove up, he noticed a polar bear inside the bear fence, looking in at the window. When the bear saw him it ran away, but not before it had nosed through a kit bag and eaten most of the foam on Steinar's snowmobile seat. Closer inspection of the bear fence found that the wire had been pulled. It looked as if the bear had trodden directly on top of the wire, but the flare hadn't fired as the trigger had become frozen in place. Tommy's unexpected arrival had been very well timed. We would have had a far ruder awakening if a bear had come through the cabin window.

After a long drive we arrived safely back in Longyearbyen. It was the first day of the midnight sun and as its rays warmed us through the window I wondered how much things would have changed by the time we returned in July.

Summer in Svalbard

The air vibrated with the buzzing wing beats of hundreds of little auks, their flocks wheeling low over the scree slopes before climbing and circling again and again. Delicate grey and white kittiwakes floated on the updrafts and scores of penguin-like guillemots sat upright in long lines, along the slender ledges and terraces of the cliff face. The Arctic summer was alive with colour and spectacle. Our crew had returned to Svalbard, chartering

Mother and cub (left): polar bears normally give birth to two cubs, but only 50 per cent of cubs will be tough enough to survive their first year.

a fishing trawler, the *MS Polarhav*, on which Martyn, Jason, and I would travel the coastline in search of bears and walruses. The coast of Svalbard was now almost entirely free of sea ice and, with the snow now melted, we could no longer use snowmobiles. The boat was the best way to access the remote corners of the coastline to look for wildlife.

The bird cliffs of Bellsund on the west coast were our first stop. Early in the morning we had ferried and hauled the endless boxes of gear, food, and fuel needed for three members of our team to stay here for two weeks. The Bellsund crew – cameraman Warwick, researcher Louise, and Arctic guide Lassa – would make their home in a small trappers' hut, not unlike our old base in Kvalvagen, to try to film the fledging of the guillemot chicks. In early July, these chicks take a brave leap from their 300m (1000ft) ledges and, accompanied by their parents, glide down to what they hope will be the safety of the sea. For many, however, this first leap will be their last. If the winds divert their course or they misjudge their glide, they will miss the water and crash into the hard ground. Remarkably most survive this brutal impact and, somewhat shell-shocked, stagger and stumble towards the sea. But it is a case of "out of the frying pan" as they are picked off by Arctic foxes, who at the peak of the fledging are almost overwhelmed by opportunities to feed their hungry cubs.

Having seen foxes in the winter, when they survive by scavenging on bear kills, the difference from this summer scene was enormous. Although it was hard to begrudge the fox cubs of Bellsund their meal, I couldn't help but root for the poor fledglings as they struggled across the tundra. Only the luckiest avoided the foxes and reached the sea. It was hard to imagine a more brutal start to life beyond the safety of the nesting ledge.

Guillemot
Uria lornvia

In the Arctic, guillemots arrive to find nest sites on narrow cliff ledges close to the sea in April to May, but often don't start laying eggs because of the snow, until June to July.

Whilst breeding, guillemots feed in open water at the edge of the summer ice, but as the ice is ever moving they may have to fly over 100km (65 miles) to feed and return to their hungry chicks. One Canadian guillemot cliff with several thousand birds was measured to consume over 1.4 million fish over the course of a single month.

When leaving the cliff-top nest, a three-week old chick leaps off and frantically flaps its way to the water below, followed closely by a concerned parent. Those that survive raft up with the adults in groups of up to several hundred and head out to sea to fledge and learn to feed on fish and crustaceans for themselves. During this swimming migration the adults moult into their winter plumage and lose their ability to fly until their flight feathers have grown out again the following season.

KEY FACTS

Location	Arctic
Habitat	sea cliffs, open ocean
Lifespan	up to 21 years
Size	29–43cm (11–17in)
Weight	700–1200g (1½–2½lb)
Food	fish, crustaceans

Black guillemots (above): less numerous than Brunnich's, they like to hunt the ice edge for fish and crustaceans, diving to depths of 75m (250ft) for up to a minute before surfacing again.

Snow geese (next page) gather in enormous flocks on route to the high arctic. Some have travelled from the southern United States to get here. Over 1.5 million geese come to Baffin Island to feed and raise their chicks.

Bird cliffs at Bellsund (right): the crew spent two weeks here, on the west coast of Svalbard's main island of Spitsbergen, in order to film the fledging of the guillemot chicks.

Arctic foxes in summer (above): two cubs play on the summer tundra. Their coats are brown and grey, typical of this time of year.

Later in the year (above): in contrast to summer, the Arctic fox's winter coat is pure white.

Arctic fox
Alopex lagopus

Arctic foxes moult twice a year, losing their thick white winter coat in May, which is replaced by a brown one in July. The winter coat begins to return in September and by December is complete. This dramatic transformation enables the foxes to cope with the seasonally changing environment.

Highly adaptable and far from fussy eaters, Arctic foxes feed opportunistically throughout the winter on carrion and carcasses left by polar bears and wolves. This nomadic lifestyle has led the fox to travel further than any other terrestrial mammal. One radio-tagged fox wandered 1000km (625 miles) in two years, travelling from Russia to Alaska. In spring and summer they patrol areas where seals, rodents, and birds gather to breed, feeding on anything from fish to berries or seal pups. They will often venture up and down cliffs to in search of accessible nests, dislodged eggs, or fallen hatchlings. So depending on the topography of the area, Arctic foxes place their den site as close to an abundant food source as possible.

KEY FACTS

Location	Arctic
Habitat	sea ice, tundra, bird cliffs.
Lifespan	up to 10 years.
Size	50–70cm (20–28in)
Weight	2.5–8kg (5½–17½lb)
Food	lemmings, eggs, carrion, berries.

As we clambered down the lower slopes of the cliff we heard the rasping call of a fox above the clamour of the thousands of seabirds. The melt had changed everything, bringing the place to life. The smells and noise of the bird cliff stood in strong contrast to the icy silence of winter.

The sea bear

We had been travelling for over a week on the Polarhav, searching for the last remaining traces of broken sea ice, known as "drift ice", the favourite summer haunt of polar bears, seals, and walruses. So far we had been unlucky, hampered by foggy low cloud and the seasickness brought on by the endless swell from the North Atlantic.

Then through the binoculars we spotted the gently curving bay of Vibebukta on the southern shore of Svalbard's Nordaustlandet Island. The bay was full of drift ice of varying densities and on the long, gently inclining shoreline we could see an amazing nine polar bears! Jason and his assistant Arne lowered the filming boat, a 4.5m (15ft) rig with an outboard motor, while Martyn and I readied the filming gear.

Jason steered the small vessel into an incredible seascape. Huge chunks of ice, infinite in their variety of shape and size, drifted past us, blue gems in the green water. The sun came out and the last of the fog lifted. We travelled through a maze of sparkling ice and water, stopping every now and then to scan the horizon for signs of life. All was quiet except for the occasional splash of a surfacing seal or the eerie distressed cries of the kittiwakes as pirating Arctic skuas mobbed them, harassing them to give up the fish in their crops.

Then Martyn spotted it – a bear slipping into the water from the distant beach. We quietly and slowly set the boat on a course to get closer. It was a female, in her prime, about six years old. We filmed as she propelled herself through the icy water with strokes of her powerful front paws, her head held just above the water, pristine white, an island of dry fur. She pulled herself out onto a piece of drift ice and shook the water from her body, twisting her head and rotating her shoulders. She sniffed the ice and the air, hunting for seals, before plopping back into the water to continue her search. The bear wasn't afraid and paid us little attention at first, then her curiosity got the better of her and she came straight for us, ducking under the water and surfacing with just her nose and eyes, peeking at us. Not wanting to get too close, we retreated and watched her from a distance. It struck me how completely adapted polar bears are to the water, how at ease this one was in her ephemeral environment, truly earning her Latin title *Ursus maritimus*, the sea bear.

Returning to the Polarhav we could see more of our surroundings now the fog had gone. We were anchored in front of a huge wall of ice, 30m (100ft) high and over 200km (125 miles) long, a line of burning white scorching the length of the horizon. This amazing feature was the Austfonna Ice Cap and here on its southern edge it dipped its toes in the summer ocean. We took the filming boat along its length as the setting sun lit up a series of spectacular waterfalls. In summer meltwater forms on the highest part of the ice cap, some 550m (1800ft) up, and carves channels in the ice, running downhill to where the ice meets the sea. Here, along the length of the white icy wall, a thousand waterfalls plummet over the edge, straight into the blue sea. Filming this phenomenon was difficult, as the

Swimming bear (above): polar bears are excellent swimmers and feel at home in the water. We even watched one that seemed to be hunting seals in open water.

A cascade of meltwater (right): the Austfonna ice cap at Brasvellbreen. Along the 100km (60 mile) wall of ice, a thousand similar water spouts empty into the ocean.

chop of the waves meant we struggled to keep the camera steady. We had to fight the dying light and the increasing ocean swell before we succeeded.

Vibebukta proved a wonderful spot for filming bears. The weather slowly changed, growing more unsettled. A brisk wind pushed the drift ice into the shore, squashing it together and making it more difficult to navigate. It was here that we saw a bear hunting one morning. We watched it lower its body into a tiny pool between two chunks of ice. It would disappear for many minutes before poking its head up again, peering low over the ice. Then suddenly it emerged, grasping a small ringed seal in its mouth. It dragged the seal backwards and forwards, circling it before biting it again and again, as if exulting in its success. Then it began to feed, tearing off the skin and long strips of steaming pink blubber. A pure white ivory gull sat at a respectful distance, waiting for scraps. Finally, having stripped the black carcass of its energy-rich blubber, the bear turned its bloody face away and headed west, disappearing in the rubbled piles of pack ice.

To the far north

The poor weather forced us on and we set a course for Svalbard's remotest island: Kvitoya, the "White Island". We crowded onto the bridge of the Polarhav as the island emerged from the fog, an ominous sight not unlike Jules Verne's lost world. Kvitoya is made almost entirely of ice, a terrifying fortress with huge 80m (250ft) ice walls rearing up out of the sea. In places the ice was overhanging, poised to fall into the water. Small chunks would tumble and crash into the sea, portents of a more catastrophic collapse. We kept a safe distance in the small filming boat. At one end on the island a pebbly spit jutted out into the sea, the only dry land around. Here we watched a painfully thin, dirty brown polar bear chewing on the open carcass of another. To turn to cannibalism he must have been desperate. With no pack ice or drift ice nearby to hunt in he was marooned on the island and the way his bones poked through his skinny hide was a reminder that while the great melt provides opportunities for many, for some it can bring disaster.

Blood on the ice (above): when a bear makes a kill, the sight of blood may act as a signal to scavengers such as Arctic skuas and the pure white ivory gull. Only a minute after we watched this bear kill a seal the first gull appeared on the scene.

"The lost world" of Kvitoya (next page): one of Svalbard's most northerly islands, the "White Island" is covered by an ice cap 410m (1350ft) deep in the centre. The ice falls gently towards the coast, where it ends abruptly with 40m (130ft) high ice cliffs.

The "rudest" animal on the planet

Travelling south again to we anchored the *Polarhav* off the southern tip of Edge Island and set off in the filming boat across open seas. Our destination was the Tusenoyane or "Thousand Islands", a scattered archipelago of low-lying rocky outcrops and a favourite haul-out site for one of the Arctic's most bizarre animals, the walrus. These giants are one of the largest members of the seal family, instantly recognizable by their huge ivory tusks.

The Tusenoyane were alive with birds. Arctic terns danced over the sheltered water, rising and falling, dipping their bills beneath the surface to snatch tiny fish. Little auks and puffins perched on the rocks, resting before they headed south. Their breeding cycle complete, their time here was coming to an end.

We steered the boat through a narrow channel between two islands, the water churning over reefs in spinning eddies. All of a sudden, on a broad beach ahead of us, the massed golden-brown bodies of 200 walruses lay in the sunshine. Not wanting to disturb their slumber we skirted around the back of the island and anchored the boat. To reach the shore we had to wade through chest-deep water, carrying the precious filming gear over our heads to keep it dry.

Martyn started filming from about a 60m (200ft) distance, keeping low and moving slowly so as not to startle the resting giants. He crept forward, gaining a few careful metres every time, until he was only 6m (20ft) away. We needn't have worried, as the walruses didn't seem in the least bothered by our presence. Walrus hunting ceased on the island in 1953, so the animals are less jumpy here than in other Arctic locations, where hunting still occurs. But our cautious approach seemed to have won the walruses' trust and they rewarded us with a remarkable insight into their behaviour.

Although they were huddled close together, a seething mass of wobbling blubber, it seemed that the walruses were reluctantly social. Arguments frequently broke out with a burst of aggressive honking, spitting, and tusk stabbing used to settle the fracas and restore the status quo. The walruses made for particularly noisy bedfellows, snoring, dribbling, belching, and farting through the afternoon, occasionally raising a paddle-like flipper to scratch at nasty-looking tubercles and patches of flaky dry skin. All the time walruses came and went, undulating their flabby bodies in and out of the shallow water. We watched a female lead her small calf past a gang of adolescent males into the sea, lashing out at any individuals who got too close, fiercely protective of her offspring.

Summer haul-out (left): tensions run high among the walruses. The protagonists rear up and show each other their tusks, honking their threat. If this doesn't settle the argument, they will stab at each other and can cause serious injury.

Walruses prefer shallow coastal waters. When the ice retreats, vast areas of coastal water are opened up for them to forage, the melt providing them with rich opportunities to feed on a huge variety of marine invertebrates. Their favourite is a bivalve clam, which they find by grazing along the sea bottom, searching and identifying prey with their sensitive whiskers, propelling themselves along with sweeps of their broad flippers. They suck the meat out by sealing the clam in their powerful lips and drawing their tongue rapidly into their mouth, piston-like, creating a sucking vacuum. These feeding habits help to stir up the sediment and increase the richness of the surrounding coastal waters still further. The walruses' love of seafood was apparent in their horrendously potent "fishy breath", which at times made our eyes water!

Walrus
Odobenus rosmarus

Walruses are huge, weighing up to 1600kg (3500lb) with males more massive overall than females. The most distinctive feature of a walrus is its tusks, which are possessed by both males and females. The largest bulls develop tusks that are over 1m (3ft) long and weigh 5kg (11lb).

Normally described as cinnamon brown in colour, walruses in fact change colour depending on their temperature. They can appear white after diving, but then turn pink when they are warm. This is because they are able to vary the blood supply to the periphery of their body so that they cool down or stay warm depending on their environment.

Their size and shape make walruses very slow on land, but in the water they are transformed. If necessary, they can dive to 100m (330ft) to retrieve clams from the seabed. A walrus can eat up to 4000 clams in one feeding. Air sacs in the walrus's neck allow it to sleep with its head held up in the water. Nursing females use this vertical position as they suckle their pups.

KEY FACTS

Location Arctic
Habitat pack ice, shore line, open water.
Lifespan up to 25 years.
Size up to 4m (13ft).
Weight up to 1600kg (3500lb).
Food clams, mussels, starfish, crabs.

Surprisingly agile (right): walruses are much more at home in the water than on land. They have been known to attack swimming polar bears.

A multipurpose tool (above): the walrus's famous tusks are used for social interaction, protection from bears, and feeding. They are also potentially useful as "ice picks", to help the animal pull up onto the rafts of sea ice. One walrus seemed particularly curious and approached the crew to take a good look.

At this time of year, walruses are forced to haul out on land, as the majority of the ice they call home has melted away. Though they can be spotted all around Svalbard in the summer months it is a continuing mystery where they spend their winters.

The walruses' behaviour entertained us long after filming had ended and we were sorry to have to load the boat up before bouncing away over the empty ocean, turned golden by the low 4am sunlight.

The greatest melt

The time had come for us to return home and we headed back towards the south cape. One last task had to be carried out before we could leave. I woke suddenly, the anchor chain whizzing past my head, on the other side of my cabin wall. We had reached our destination, Kvalvagen, the site of our winter adventure. Here we hoped to complete the lapsed time positions that we had set up on our previous trip.

The change in these familiar surroundings was startling. The small hut that we had frequently had to dig out from under the drifting snow stood in a field of green grass and sedges. The mountains behind were a desert of sunburnt rock. With the snow gone, no snowmobile could operate and we had to travel on foot, keeping our eyes peeled for bears. The bird ledge where we had watched Scar Nose wander across a sea frozen right to the horizon was now pounded by ocean swell.

The glacier fronts in Kvalvagen, once locked in by fast ice, were crumbling and falling into the shallow sea. Where a few months ago we had walked on flat expanses of shimmering sea ice or snowmobiled over it in pursuit of polar bears, we now floated, riding the boat between the waves. These were places we had come to know well, and while they were recognizable they seemed forever altered, as if our memories of them were nothing but a dream.

On returning home I was saddened and shocked to read that scientists had announced that the summer of 2007 had seen the greatest great melt ever recorded. Only 4.13 million sq km (1.6 million square miles) of ice remained. The sea ice had never retreated so far or so fast. For the first time, the fabled North-west Passage – the shipping route from Europe to Asia through the Canadian Arctic – was completely free of ice. Climate change is thought to be the culprit: each year it seems an additional 100,000sq km (40,000 square miles) of ice is lost, but in 2007 the rate of loss leapt to a staggering 1 million sq km (400,000 square miles).

The most dramatic predictions have suggested that the Arctic might be completely ice free as soon as 2013, just five or six summers away. The future of the walrus, the millions of seabirds, the whales, the foxes, and the king of the Arctic, the polar bear, would appear to hang in the balance. Unless we act now, we stand to lose this most wondrous of Nature's Great Events.

Last days of the melt (above): the late summer sun sinks closer to the horizon over the needle-sharp mountains of Hornsund on the west coast of Spitsbergen Island.

The Great Salmon Run
British Columbia

The life cycle of the Pacific salmon is one of the classic tragedies of the animal kingdom. After surviving years at sea avoiding predators and growing to adult size, then swimming thousands of miles to return to the rivers in which they were born, struggling upstream against strong currents and the bears, wolves, and eagles waiting to eat them, they lay their eggs or sperm, and then die. It is an epic journey that always ends in the same way – with death. But from this heroic adventure, new life is born and the trials and tribulations of another generation of salmon begin.

The return of the Pacific salmon each autumn to the terrestrial landscape surrounding the North Pacific is one of the greatest natural events on the planet. It happens on a scale virtually unrivalled in the animal kingdom and is the defining annual event in the lives of every animal that lives here. Nowhere is this more evident than in the coastal forests along the north-western coastline of North America, the largest area of intact temperate rainforest in the world and one of the – if not the – most productive landscape on the planet. The rivers and their innumerable tributaries spread through the entire length and breadth of the coastal region, carrying with them the salmon that provides the fuel to sustain the entire ecosystem. The ocean, the rivers, and the land conspire to produce the perfect conditions for life.

British Columbia
Canada

British Columbia's coastal temperate rainforest is characterized by some of the oldest and largest trees on Earth, the most common of which include Sitka spruce, red cedar, and Douglas fir. Trees can tower up to 300 feet and grow for more than 1,500 years.

The biological abundance of British Columbia's coastal rainforests are the result of over 10,000 years of evolution that began when the glaciers of the Pleistocene epoch melted. This forest used to stretch continuously along the coast from California to Alaska. Today nearly 60 per cent of the world's coastal temperate rainforests have been logged or developed. The Great Rainforest represents one-quarter of what remains.

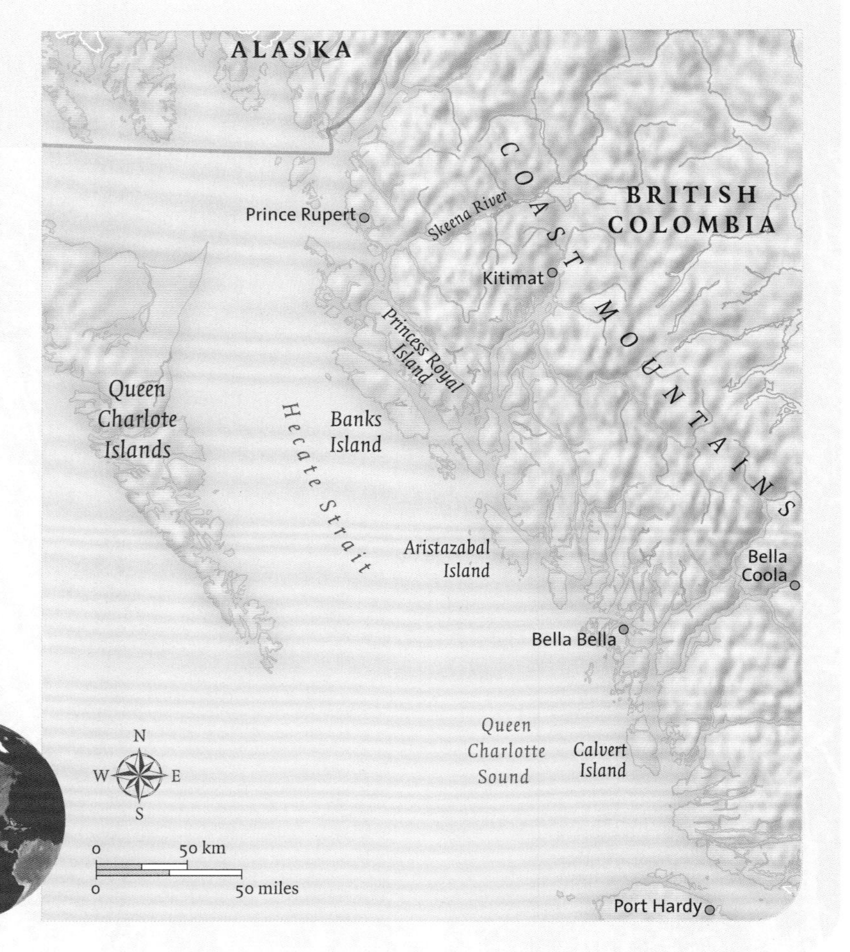

Khutzeymateen River (right): on the north coast of British Columbia, this is a classic temperate rainforest river system. Where the river enters the ocean, at the end of a long and narrow inlet, a large grassy estuary is formed. Old growth temperate forests cover the valley and inlet walls, while snowy mountain peaks tower above the inlet and river.

A unique habitat

The North Pacific Ocean is a huge area (about 85 million sq km/33 million square miles), with a shoreline stretching north from the islands of Japan, along the Russian Far East coast, across the Bering Sea to Alaska and down the west coast of North America to California – a distance of over 15,000km (10,000 miles). Here there is a unique relationship between ocean and land. It is a relationship that is advanced by the Pacific salmon, whose life cycle is one of the defining characteristics of this amazing part of the planet.

Lying in the northern hemisphere between the 49 and 60 degrees north, the temperate rainforest is smack in the middle of the Earth's northern temperate zone. Most of the forest is in the boundaries of Canada, but it extends into the southern coast of the Alaska panhandle before

The heart of the Great Forest (above): what is left of this ancient forest runs along the British Columbia coast from Vancouver all the way to the Alaska border and beyond, an area roughly 800km by 150km (500 by 100 miles).

petering out near the border with the Yukon Territories. As it stretches north, it begins to be compromised by more arctic conditions – colder temperatures, less sunlight, and shorter growing seasons. The forest itself is teeming with life. Although the rainforests at temperate latitudes do not have the great diversity of life that is found in their tropical counterparts, the massive nature of the trees here and their physical size and number combine to tip the balance of life per square metre in favour of temperate rainforests. The understorey comprises mainly shrubs, ferns, and mosses. Flowering herbs and grass that need more sunlight just can't compete under the darkness of the forest canopy. The forest floor is carpeted in moss. Stream sides and damp areas host ferns and wherever breaks in the canopy allow sunlight in – along streams or rocky openings – shrubs such

Teeming with life (above): the old growth temperate rainforest of coastal British Columbia and Alaska contains more life in terms of kilograms per square metre than tropical rainforests.

as salmon berry, rose, blueberry, elderberry, mountain ash, and devil's club provide important summer food for the grizzly and black bears.

Nursery logs (above): not all the plant life in the forest grows on the ground. An old Sitka spruce can carry several hundred kilograms of plant life on its broad trunk and branches.

The forest trees

Not all the plant life in the forest grows on the ground. The trees themselves carry heavy loads of mosses, ferns, and even some shrubs and small trees. An old Sitka spruce can carry several hundred kilograms of plant life on its broad trunk and branches. And when an old tree falls in the forest its broken body provides a good rooting environment for new ones. The soil in the forest can be quite barren and washed of nutrients and minerals, so a seed from a nearby tree will find more to sustain it on the fallen trunk of one of its neighbours than on the forest floor. These fallen giants are known as nursery logs and several trees may grow along their length as they decay and are absorbed back into the soil.

Centuries ago, the temperate rainforests would have extended many hundreds of kilometres further south into the continental United States. Along the coastline of Washington, Oregon, and into northern California the pristine relationship between salmon and forest would have continued but for the influence of man. But the great coastal forests are largely gone, consumed by the voracious appetites of the timber industry and the salmon runs so overfished that they are a shadow of what they once were.

Today the heart of the Great Forest is along the British Columbia coast from Vancouver all the way to the Alaska border and beyond, an area roughly 800km by 150km (500 by 100 miles) that includes thousands of islands and a massive mountain chain broken by hundreds of fjords and inlets to produce a coastline more than 1000 times longer than its actual length.

Here the Coast Mountains reach down to the sea. Steep, granite-sided fjords cut deep into them. Large rivers and watersheds reach 20, 30, 50km (12, 18, 30 miles) inland. In the valley bottoms of the larger river systems, the forest grows to its maximum size. This is where you find the giants – Sitka spruce 2m (6ft 6 in) in diameter and hundreds of years old; ancient red cedar, many still bearing scars from their use by native people centuries ago; and western hemlock and yellow cedar 45m (150ft) and taller.

Giants of the forest (right): in the valley bottoms of the larger river systems the forest grows to its maximum size. This is where you find such monsters as the Sitka spruce, 2m (6ft 6 in) and more in diameter, and hundreds of years old.

The salmon people

The ancestors of the First Nations people taught that there was a spirit in all things – the plants, the animals, the fish, and the birds – and that a little of that spirit was taken from everything they ate. So intertwined with nature and its gift, the people of British Columbia's coast reflect the link between land and sea. Central to this relationship are salmon, hence their title the "salmon people".

A dietary staple even in modern society, salmon is also the lifeblood of the economy for many First Nations groups living on the coast. Village residents roast fresh sockeye in the old way, over an open fire, and fill their pantries with hand-canned fish prepared according to countless different recipes. In some areas, for instance Bella Bella on the central coast of British Columbia, workers pack wild chum salmon roe for export to Japan.

Much like contemporary Christmas or New Year's celebrations, aboriginal First Salmon ceremonies recognized the ever-revolving circle of life. With supplies of dried, smoked, and canned salmon running short after a long winter, the arrival of the first fresh fish of the season was eagerly awaited and gratefully acknowledged. Throughout the Pacific Northwest, the Tlingit, the Haida, the Tsimsyan, the Heiltsuk, and many other tribes ceremonially burned the bones of the fish, or returned them solemnly to the stream in which they had been caught. The bones were believed to find their way back to the village of the salmon people, where the first salmon would be whole again, thus assuring a plentiful food supply for the following year.

Totem carving (above): this ancient practice has been handed down through generations as a way of preserving the history of local native heritage as well as honouring tribal rituals and sacred spirits of people.

The lifeblood of the Great Forest

From the headwaters in the glaciers and ice fields of the mountains the water flows clean and pure, sometimes hundreds of kilometers, to empty into the ocean. The forest is essential to preserving the quality of the river water and providing a home for the returning salmon. The presence of trees along the banks of the river regulates temperature levels in the water, providing for the development of eggs over winter and the survival of young salmon when they emerge in the spring. The overhanging trees and vegetation provide leaf litter, insects, and other sources of food for young salmon. Old trees that fall into streams alter the flow of water and the structure of the stream bed, creating both spawning and rearing habitat. And finally the strong, deep web of roots that extend through the soil helps to anchor the life-giving soil to the hard pan of granite beneath, ensuring that sediment does not wash into the streams in the heavy winter rains, smothering and killing the developing eggs sheltered there.

The rivers are the watery link between the land and the sea and are what allows this productive forest to survive. It is through this link that the salmon are able to leave their saltwater, ocean environment and enter the freshwater, forest one. The sheer weight of living tissue contained within the salmon that leave the ocean to travel up the freshwater arteries into the heart of the forest can be measured in millions of tonnes. The forest is like a living organism – rivers, creeks, and streams are its arteries and veins, carrying the nutrient-rich blood and flesh of the salmon. And the analogy to a living body is more than a metaphor. The salmon are literally the lifeblood that feeds not only the inhabitants of the forest – more than 200 species sustain themselves on salmon – but also the forest itself.

Researchers have sampled and measured the composition of the living tissue of trees throughout the Great Forest and have discovered marine nutrients: nitrogen and carbon isotopes that have their origins in the sea. The North Pacific is one of the richest and most productive oceans in the world. Many of these rich marine nutrients are finding their way dozens of kilometers inland to the Great Forest, and it is all down to the salmon.

But getting back to their forest-river homes is no easy feat for the salmon. It's one of the most difficult and dangerous journeys in the animal kingdom. They have to overcome many obstacles to reach their goal, which they do with a relentless effort. First they must run a gauntlet of at least 1000km (625 mile) of marine-based predators trying to eat them before they even get to the rivers. Then they have to battle their way upstream, sometimes hundreds of kilometres, against fast-flowing water where every metre of river takes more of their dwindling energy reserves. All along the river hordes of hungry birds and mammals wait to pounce whenever the salmon cross a stretch of shallow water. Finally, there are the waterfalls and

An extraordinary bond (left): the productivity of the temperate rainforest of British Columbia relies on the unique relationship between the rich waters of the North Pacific Ocean and the coastal mountains and islands.

Pacific salmon

Pacific salmon are an "anadromous" fish. Anadromous means "up-running", and anadromous species hatch and live the first part of their lives in fresh water, then migrate to the ocean to spend their adult lives, finally returning to fresh water to spawn. Pacific salmon spend up to eight years out in the Pacific Ocean, and, when they reach sexual maturity, they return to the freshwater stream of their origin to lay their eggs. They are unique in that they make the round trip out to the ocean and back to their natal streams to spawn just once: after spawning they die.

There are five known species of Pacific salmon: chum, coho, pink, sockeye, and chinook. Although many rivers and streams in British Columbia provide breeding grounds for more than one species, every freshwater system has its own genetically unique salmon populations, which would be unable to interbreed successfully with populations from other systems.

Chum salmon *(Oncorhynchus keta)* have the widest natural geographic distribution of all Pacific salmon species, ranging from Korea to the Arctic coasts of Russia and Alaska and as far south as Monterey, California. They're also the second largest species in the Pacific (adults can reach 20kg/44lb). Chum spawn successfully in streams of various sizes, and the fry migrate directly to the sea soon after emergence. After distributing themselves widely throughout the North Pacific Ocean, the fry return to their home streams at various ages, usually between two and five years old, but sometimes as old as seven.

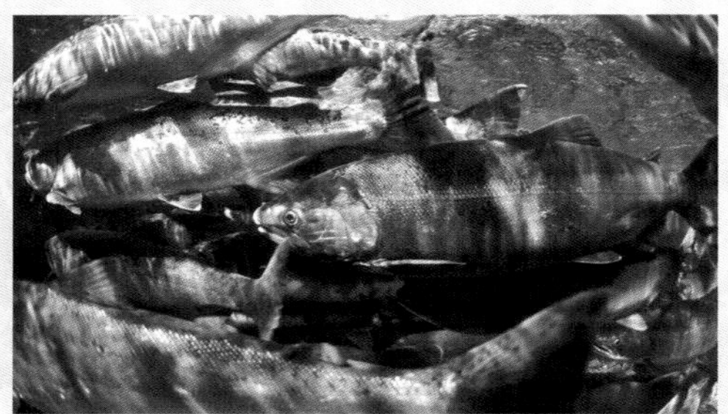

Coho salmon *(Oncorhynchus kisutch)* are also widely distributed throughout the North Pacific, though they occur in smaller numbers than other Pacific salmon species. The majority mature during summer in their third year of life, having spent four to six months in incubation and up to 15 months rearing in fresh water followed by a 16-month growing period in sea water. Coho are bigger than pink salmon (see right) – 2–4.5kg (4–10lb) on average, though 9kg (20lb) is not uncommon – and are distinguishable by the presence of a square, silver patch with a few scattered spots on the upper portion of the tail. Coho are also much more powerful swimmers than pinks and can negotiate their way over waterfalls and rapids that will stop pink salmon.

Pink salmon *(Oncorhynchus gorbuscha)* are the most abundant of the five species, and are distinguished from Pacific salmon by a number of features. As adults they are the smallest (averaging 1–2.5kg/2–5lb), and when returning to their natal stream to spawn, males develop a remarkably large hump on their backs. The reason for this is not clear, though some scientists believe it may be designed to divert the attention of predators away from the females. Both sexes change colour when they migrate upstream, from bright silver to dark grey on the back with a white to yellowish belly. The fry migrate quickly to the sea after emergence and grow rapidly as they make extensive feeding migrations. Within two years they return to their home stream to spawn and die.

Sockeye salmon *(Oncorhynchus nerka)* are the most spectacularly coloured of all Pacific salmon, and for this reason possibly the best known. According to W. E. Ricker of the American Fisheries Society, "sockeye" is a corruption of the name used by First Nations people of southern British Columbia, originally printed as "sukkai". Sockeye salmon exhibit a greater variety of life history patterns than the other species, and characteristically make more use of lakes as a habitat in their juvenile stages. Although sockeye are primarily anadromous, there are distinct populations called kokanee that mature, spawn, and die in fresh water without a period of life at sea. Those that do migrate to sea do so soon after spending a year in a freshwater lake after emergence, and spend from one to four years there before returning to spawn.

Chinook salmon *(Oncorhynchus tshawytscha)* can be distinguished from other Pacific species by their large size (adults can reach 45kg/100lb), small black spots on both lobes of the caudal (tail) fin, and black pigment along the base of the teeth. They are also known to have a large number of pyloric caeca in their gut to help regulation of salt in their bodies when going between fresh and salt water. "Ocean-type" chinook are typical of populations on the British Columbian coast, which migrate to sea during their first year of life, normally within three months after emergence, spend most of their lives in coastal waters, and return to their natal river in the autumn a few days or weeks before spawning.

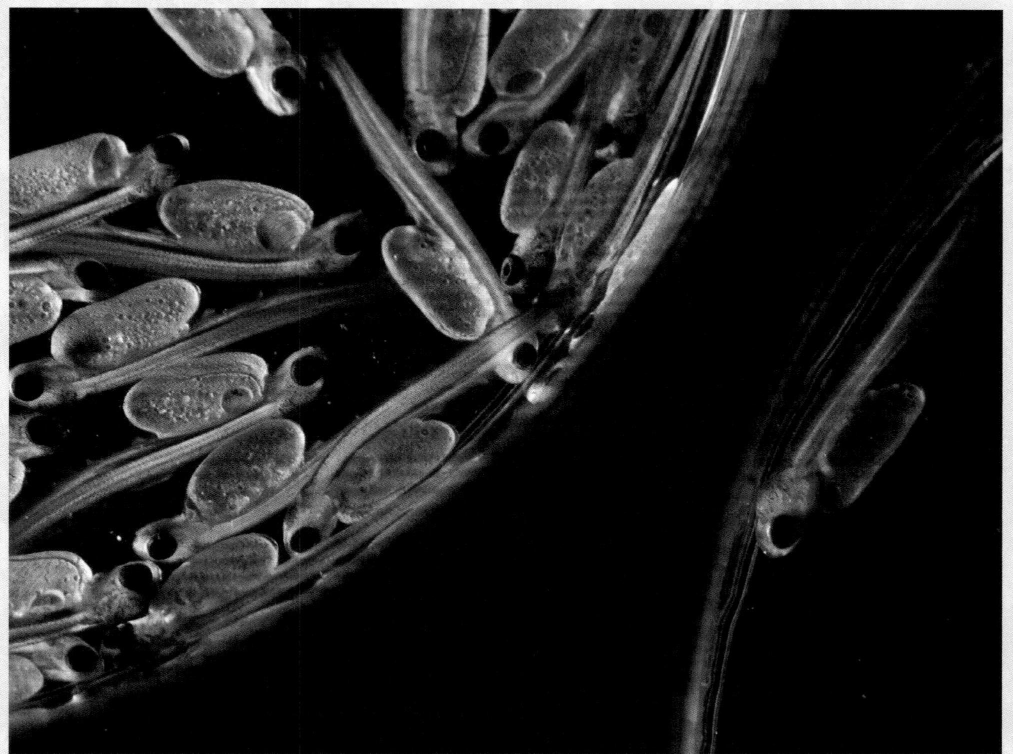

Small fry

Though there are exceptions, most populations of Pacific salmon follow the same general life cycle. Eggs are deposited and buried in gravel nests called redds. The size of gravel chosen for the nest depends on the size of the female parent. The embryos incubate within an egg membrane for several months. After hatching they become known as alevins, and are attached to a large, external yolk sac for nourishment. Once the yolk sac has been metabolized by the alevin, the young salmon can feed on its own. It then wriggles up through the gravel in spring and emerges as fry. The fry can remain in fresh water streams or lakes for anything from just a few hours to two years before swimming out to the ocean. Those remaining in the fresh water become known as parr because of the vertical brown-green bars that develop on their sides, providing camouflage. Once at sea, the species undertake migrations of varying distance, lasting up to several years.

rapids, often the hardest obstacles to overcome. Where these are too steep to swim, where the current is just too powerful, the salmon must leap 3, 4, 5m (10, 13, 16ft) or more if they are to reach their goal. I have watched them come to falls that it seemed impossible they could ever get over. And yet they do: from somewhere, somehow, mustering the energy they need. The instinct to reach their goal is so strong that it appears as if nothing can stop them. It is easy to see the heroic in their efforts.

The network of waterways that infiltrates the Great Forest is extensive, but still, it does not touch every part. Salmon are water creatures, their lives and deaths confined by aquatic boundaries. When they die many of their bodies are washed back down the rivers, returning their rich nutrient loads to the sea. Only the few trees that live along the edges of the rivers, those whose roots might extend into and beneath that watery environment, would benefit from the salmon's journey. The forest would be great, but only in a thin margin along a river's edge – if there were no bears or wolves.

Forest carnivores

The animals that feed extensively on the salmon are key players in the life cycle of the forest. Bears and wolves do not spend all of their time catching and eating fish in the rivers. Large and dominant bears can usually eat wherever they choose and will often seek the shelter of the forest to have their meal in peace, knowing other bears that might be fishing in the area will not disturb them. Wolves will do the same. Even large dominant bears,

Brooks Falls (left): this part of Katmai National Park is one of the most famous bear fishing spots in North America. Every year thousands of people fly into this remote location to watch the bears catch salmon as the fish leap over this 2m (6ft 6in) waterfall barrier.

Coastal wolf (above): an adult peers warily through the long grass beside an estuary. These are enigmatic creatures – many people in the area are not even aware that they exist.

Waiting for the salmon (above): river otters (*Lutra canadensis*) are common through the temperate rainforest. They are one of the many creatures in the forest for which salmon are a mainstay of the diet.

when they are ready to leave the forest and find a quiet spot to rest, will take a salmon with them – one for the road. In one study it was determined that during six weeks of a salmon run, eight bears had transported 3000 salmon into the forest along a single-kilometre stretch of river. That would amount to somewhere between 6800 and 20,400kg (15,000 and 45,000lb) of salmon being moved out of the aquatic environment of the river and into the terrestrial environment of the forest.

Not all the salmon is consumed by bears. When the run is at its peak and the streams are overflowing with fish, bears have a glut of choices. They will eat only the richest parts of the salmon – skin, brains, and eggs. On average throughout the season, a bear might consume half of the salmon it catches. Scavengers and insects consume a small percentage more. Consequently along a single-kilometre stretch of a salmon river there could be several tonnes of salmon carcasses left to decay on the forest floor. As the salmon decays the marine nutrients locked within its tissues are released into the soil and taken up by the roots of the trees. In measuring the growth of trees in the forest, the annual growth is two and a half times greater where they have access to salmon than where they do not.

Other salmon-eating mammals, not to mention eagles, ravens, gulls, and other smaller birds, contribute to spreading salmon tissue throughout the forest. The birds' droppings contain concentrations of marine nitrogen that helps spread this nutrient throughout the forest. Mammals such as mink and otters also transport salmon into the forest away from the creeks.

Many species of flies and beetles feed on salmon carcasses washed up along the river banks. These insects, fattened on marine nutrients, are in turn preyed upon by various species of song birds that make the forest their home – swallows, thrushes, chickadees, fly catchers, vireos, warblers, sparrows, and such. And again the birds' droppings fertilize the forest. At first it might seem that this transfer of nutrients could not amount to much, but consider the volume of insects involved. One study done on the coast of British Columbia found that every 5m (16ft) along a spawning river produced 1kg (2.2lb) of fly larvae. This means that the larvae available for consumption along salmon streams can also be measured in tonnes.

The remains of the feast (left): it's always easy to tell when there are wolves fishing along a forest stream. Coastal wolves are very particular in how they eat the salmon they catch: they will only consume the brains and/or head.

Metamorphosis (next page): all species of Pacific salmon undergo some form of physical change for breeding. Sockeyes are the most colourful. Their colour changes from the bright silver of their ocean life to the brilliant red body and green head of their breeding cycle.

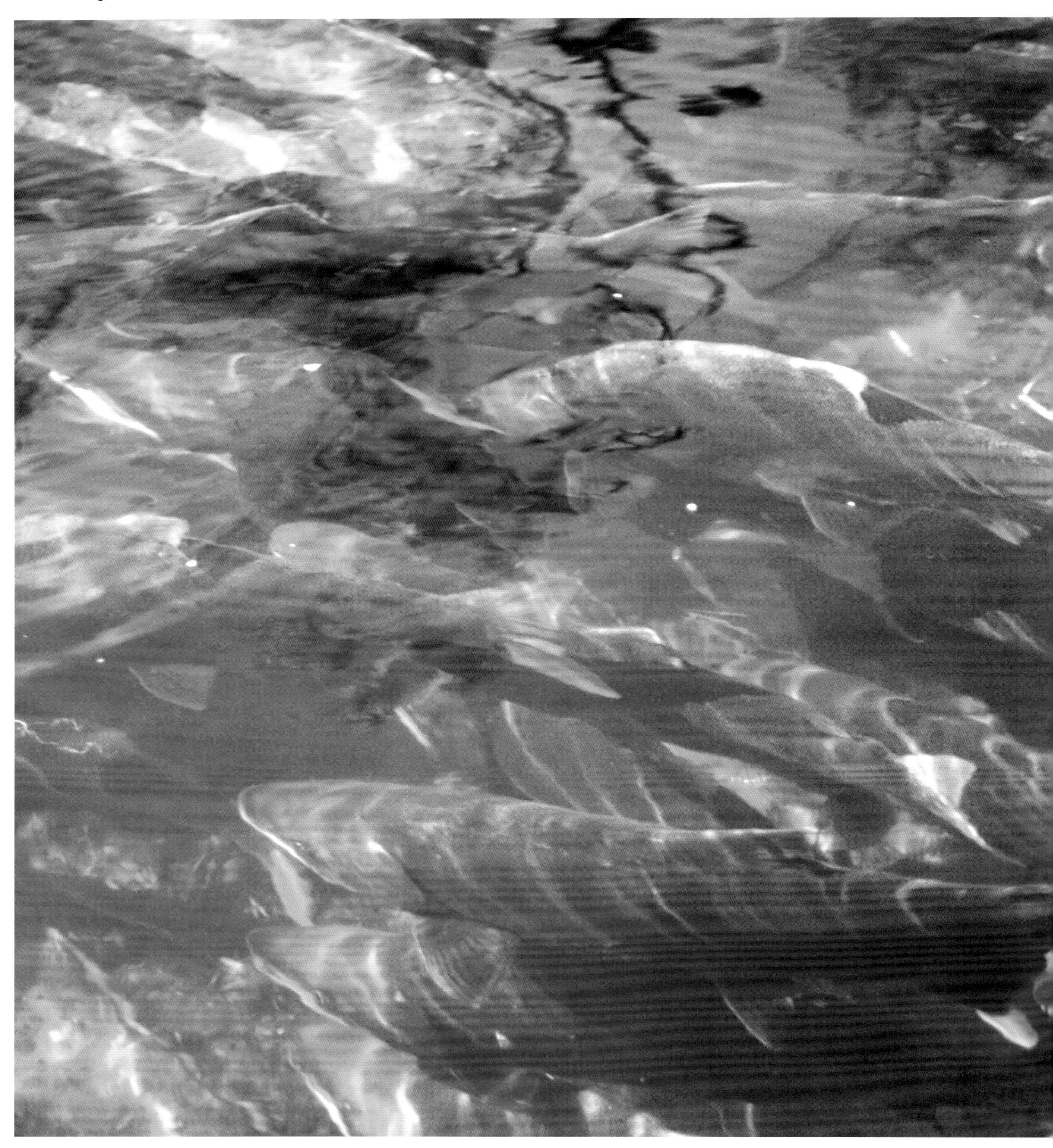

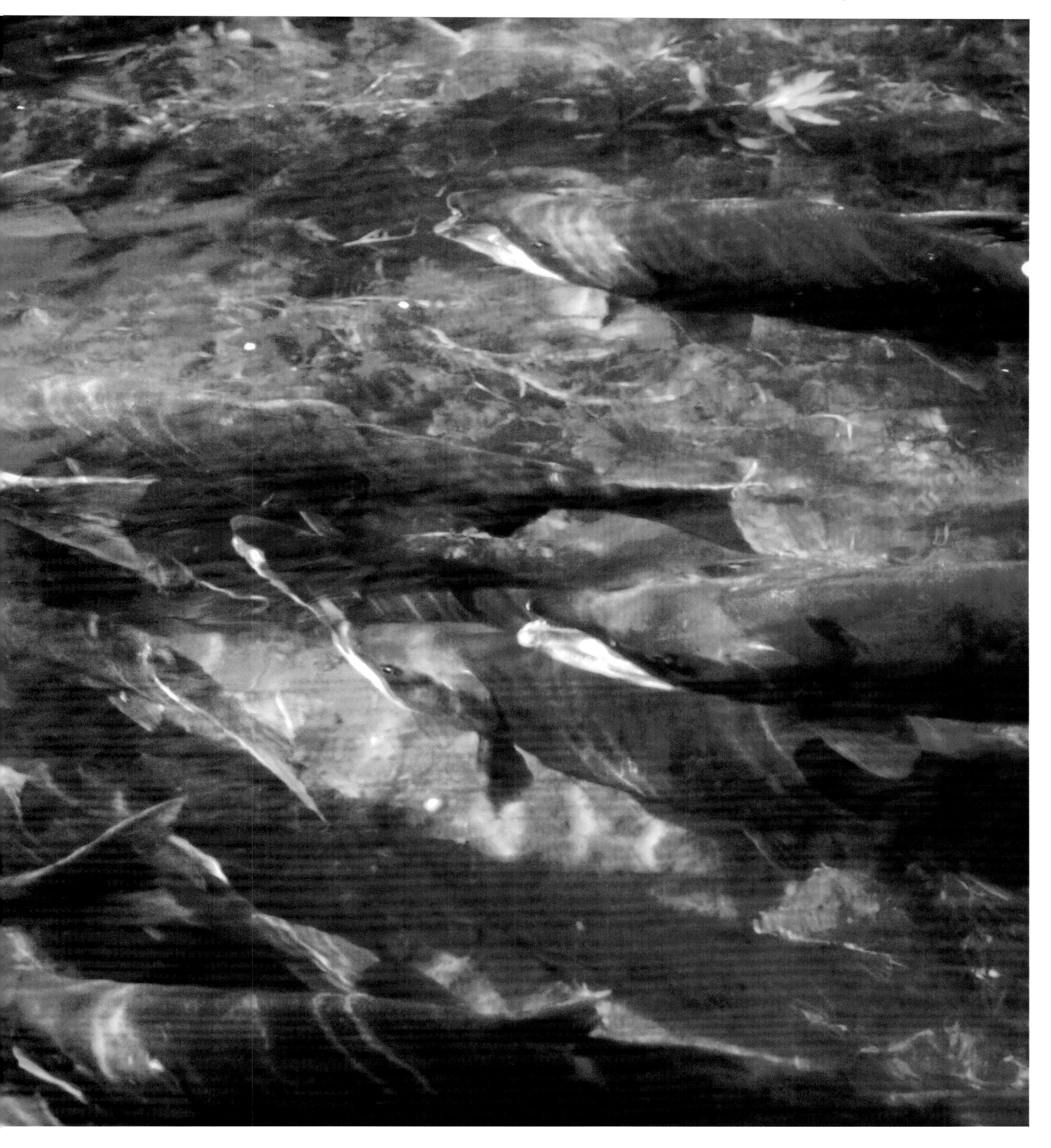

In the Great Forest
Jeff Turner

It is the heroic aspects of the salmon that have always drawn me to them. I wanted to be able to tell the whole story of their amazing life cycle and some of the major animals in the Great Forest whose lives revolve around them. But before I could do that there was a lot of work to do. Natural history films require a huge amount of preparation. Research, scripting, and planning happen for months and sometimes years before a single camera is switched on. First we had to find out as much as possible about our subjects.

There are five species of Pacific salmon – sockeye, pink, coho, chum, and chinook – and they are found throughout the Great Forest. The distinguishing element of a Pacific salmon's life cycle is that it ends after it reproduces. Atlantic salmon and other species of trout that venture between the sea and fresh water live to reproduce several times. The life cycle of the Pacific salmon is the same across all species and varies only in the timing of when they go to sea as juveniles and when they return to spawn.

When I was in England beginning to prepare for the production of this film, millions upon millions of fertilized salmon eggs were slowly developing, nestled amongst the rocks and gravels in the cold, clear water of the numerous rivers of the Great Forest. By early spring the tiny salmon – called fry – less than 2.5cm (1in) long, would be born. Most head straight down the river and out into the sea. Some will spend a year or so in the river first, but the fry of one species, the sockeye, emerge from the gravel, then swim to the nearest lake to spend the next year. What's really interesting about sockeye is that the babies, just emerging from the gravel, know where the nearest lake is even if it's upstream. They do not simply get swept downstream to the first suitable rearing habitat. They know where to go even if it means swimming upriver against the current – though they have just come into existence with no experience of the world and there is no one

The long journey begins (left): Pacific salmon enter fresh water on their way back to their spawning grounds. They have many obstacles to overcome – predators, and hundreds of kilometres of struggling upstream against currents and waterfalls.

to show them the way. How do they know to do this? As yet science has no answers.

Once out to sea the young salmon have a dangerous and difficult life ahead of them to reach adulthood. The ocean is filled with food and the salmon can quickly grow, as long as they can avoid becoming food themselves. They have to eat without being eaten. Not many of them make it. For each thousand of fry that head to the sea each spring, only a handful will return to spawn. Sharks, killer whales, seals, sea lions, and any fish big enough to fit one in its mouth all feed on salmon.

Far out in the middle of the North Pacific is where the salmon head. How they live and travel, and what sort of social grouping exists when they are out to sea are still mysteries. But at some point, after spending two or more years out in middle of this vast ocean, something tells them it's time to go home. No one knows for sure what the signal is or how it's transmitted, but vast schools of fish begin to head for the streams and rivers in which they were born. They have a long way to swim – hundreds if not thousands of kilometers of ocean to cross. By the time a pink salmon returns to its natal stream it will have swum up to 7500km (5000 miles) and a sockeye over twice that distance.

Into the mountains

In April, while the salmon were just beginning their epic journey back towards the coast of North America from far out in the middle of the North Pacific, I was beginning the first filming trip for this project, jumping on a plane in Vancouver for a flight to Alaska. We would be filming a long way from where the salmon were, but the beginning of each year sees our main characters spread out across the landscape – it is the arrival of the salmon that will draw them together later in the year. This couldn't be more obvious than with the grizzly bear. It begins its year emerging from a den high in the snow-covered mountains of the coastal ranges. Filming grizzly bears emerging from the den is one of the great natural sequences yet to be caught on film. I have tried it on many occasions, investing lots of time and money over the years, but have never succeeded. When we started the film about the salmon run for *Nature's Great Events*, we knew that grizzly bears were going to be one of our key characters. It was something I had to try again.

We located an area at the southern end of the Alaska Range that had a high concentration of denning bears – up to 50 different den sites. All we needed was one, but finding just the right one is incredibly difficult. First the den has to belong to a mother with cubs – the younger the better.

Where the rivers meet the sea (left): hundreds of thousands of tiny salmon fry enter the oceans each spring to begin their life there. For some of them it will be as much as seven years before they are ready to return to spawn.

Life on the move (above): sockeye salmon spend their first year in a freshwater lake, then three years in the ocean before returning to their spawning grounds. Some will have swum 15,000km (over 9000 miles) by the time they return to their natal stream.

That's because a mother with cubs tends to spend a good week or so around her den when she first emerges in the spring. Single bears will often just come out from the winter den and walk away, so the chances of finding one at exactly the right moment are impossibly small.

Next, the terrain is very steep and rugged. Often a den is located in such a way that you can't get close enough to it on the ground to get a good filming angle. Sometimes the ground is approachable but there are no terrain features, such as a gully or ridge, to use for cover so that you can get close enough to film without disturbing the animals. A lot of things have to be lined up just right in order for this to work.

Most grizzly bears move high up into the mountains to find places to den. They will dig into the steep hillsides in the fall before the snow comes, burrowing 2–3m (6–10ft) into the slope, and then excavate a small chamber just big enough for their curled body to sleep in. Most of these dens are on the north- or east-facing slopes of mountains where the snow tends to blow deepest and stay the longest in the spring. The snow is what insulates the bears from the freezing air temperatures of the northern winters.

Mother bears often den in the highest and most barren of places. There is nothing to eat in these areas in the spring and this is the main reason the females choose them: they are trying to stay away from other bears. In most populations of grizzly bears, some males will make a habit of killing cubs for food. Young cubs, just out of the den, are a tempting and easy target. The only defence the mothers have is to try hide in the mountains.

Scouting for bears

After flying and searching for about eight days, we had seen more than 30 different bears in the mountains and found over a dozen den sites, but nothing with the right combination of bears and terrain. Then on our

last day, high up on a very steep slope below some rugged cliffs, we spotted a mother and a yearling cub sitting in their den entrance. It wasn't an ideal situation. The den was on a steep slope above the glacier and the approach would be tricky, but I thought we could probably hike there without mountaineering equipment. I tried to pick out the shortest and safest route, but it's always hard to tell from the air how things will look on the ground, and it's even harder to judge distances accurately.

We quickly flew back to the airport and called in a helicopter. A couple of hours later we landed in a basin near the top of the mountain. Paul Zakora, my camera assistant, and I started out with our heavy packs and snowshoes on what I thought would be about a 20-minute hike to get into position. Instead it turned into a three-hour major expedition. First the light conditions went to hell. The late afternoon sun was lowering and a bank of high but heavy cloud had slid in like a huge diffusion filter, rendering the light flat and featureless – so much so that it did not reveal any detail in the white landscape. As we were snowshoeing across the basin, all the snow around me blended into a seamless haze of white.

I couldn't tell whether the ground immediately in front of me went up or down or sloped to the right or left. If I stared at the ground just in front of my snowshoes I could make out a slight texture to the snow, enough so that I could feel where my next step was going to make contact with it, but beyond that it was a mystery. I would take a step thinking I was going down a gentle slope, only to find my foot hitting the ground unexpectedly early as I had reached the bottom of the slope and was now facing a slight rise. It was form of snow-blindness. I have lived in the mountains all my life, but I had never experienced anything like the conditions I was in now.

We were making agonizingly slow progress. Paul then remembered a trick his mountaineering brother had told him about. Making a snowball,

The Coast Mountains (above): the close proximity of these mountains to the ocean is one of the reasons for the wet, temperate conditions that contribute to the productivity of the region's forests.

he rolled it out in front of himself like a bowling ball. You could follow the course of the snowball for about 5m (16ft) before it was lost in the white haze, but it was enough to tell us which way the ground was sloping.

With this new technique we finally reached the bottom of the gully. It had taken us almost an hour to cover 1km (⅔ mile). Our packs with camera gear and tripods and everything else weighed about 30kg (65lb) each. It was the first filming trip of the season and I was sweating profusely from the effort.

From the air it had looked as if this gully would give us easy access to the ridge above the den. But it turned out to be a lot steeper than I had expected. It must have been a 50-degree slope up between the rocks. Standing, I could just about reach out my arm and touch the snow at shoulder height. There was no way we could get up this with snowshoes on, so we had to sling them onto the top of our already overloaded packs. Then we started the laborious task of kicking footholds into the snow in front of us. We had no ice axes or shovels, only our winter boots and gloves to keep a grip on the steep slope. It took us the better part of another hour, but then it looked as if we had only about 300m (1000ft) across a fairly steep snowy side hill to reach the rocky knoll from which we should be able to see the den.

After a brief rest we got up, and, shouldering our packs, started out on snowshoes again. I was in the lead and had snowball-rolling duty. The going was easy, the hillside a steep but smooth slope across to the knoll. The snowball would roll straight out in front of me about 4m (13ft) and then curve gently down the slope to my right before I could no longer distinguish it from the snowy ground it was rolling along. I covered those few metres, then picked up a handful of snow and made another snowball. I rolled it out in front of me and about 2m (6ft 6in) in front of my snowshoes it disappeared. Another snowball and the same thing – gone after 2m (6ft 6in). It was as if it had silently passed a threshold into another dimension. It was eerie. There was no sound. Nothing. It just vanished.

Then it hit me. The ground between the ridge and where we were standing was not a steep-sided hill. We were standing on a cliff! But how far down did it go? I inched ahead and tossed another snowball, hoping to see it land so that I could get some sense of the drop ahead – but nothing. It just disappeared over the white edge.

That was as far as we could go. We had no ropes, crampons, ice axes, or safety equipment to help us move across this mountain landscape. It was impossible to tell how far either up or down we would have to go to get around the snowy ledge we were on and it was too dangerous to go poking about in these conditions. This was as close as we were going to get this year. I knew, however, that I had one more spring in which to film and that I was going to be coming back here to try my luck again next year.

To the sea

Although we began our filming in the high coastal mountains, the majority of our shooting time would be spent at sea level. The old-growth temperate rainforests of British Columbia were an area I knew well. I had made a few films here and, as it is one of the most beautiful and spectacular landscapes on the planet, I was eager to return.

The main characters in the film were going to be grizzly bears, wolves, and of course the salmon. But the salmon don't begin to arrive until early August, so what do the wolves and bears do until then? We wanted to observe and capture their behaviour as they went through the spring and early summer. It must be a difficult and stressful time for them as they subsist on meagre rations, waiting for the abundance of protein-rich salmon to fill the forest rivers and streams. But filming two of North America's largest predators was not going to be easy.

These two top predators are totally different characters and require different approaches to their filming. Bears are curious and interested in everything in the world around them. They will investigate anything that might yield a potential food source. They are quite fearless, too. Having been the largest and toughest in the animal kingdom of North America since the last ice age, they have learned that there is little in the woods that they need to fear. This makes them naturally willing to explore new circumstances and situations. The wolf, on the other hand, is by its very nature cautious. Wolves survive by hunting; a physically demanding task. By working together as a team, they bring down and kill prey that is often many times larger than they are. Being hurt in this process is a constant threat. If a wolf is injured, even a broken leg, its chance of surviving is slim. It can't afford to take unnecessary risks. So when wolves encounter new circumstances and situations they proceed slowly and are easily spooked. That's why it's unusual for wolves to kill animals they haven't learned to hunt as prey, and perhaps one of the reasons why wolf attacks on humans are so rare.

As I left Alaska in early May after my failed attempt to film grizzly bears, I was on the same journey as the salmon heading towards the central coast of British Columbia. Flying over the numerous green islands and intersecting blue waterways along the coast, I wondered where the salmon were now. It was still a couple months before they would reach these inshore waters.

No sign of the wolves

If I thought that filming grizzly bears coming out of their dens in the mountains was going to be hard, finding and filming wolves at their dens was going to make that look like a cake walk. Wolves are notoriously secretive about their den sites.

It was a daunting task to think about heading out into 120,000sq km (46,000 square miles) of uninhabited wilderness looking for wolves and bears to film. Filming animals in forests is one of the most difficult things to do in this business. Lines of sight are short and it's impossible to move without alerting the animals to your presence. However, we had two things going for us. The first was that in the spring, grizzly bears move out of the forest and spend time feeding on plants in the open grassy estuaries where the rivers enter the sea. Second, our best chance of filming wolves in the spring lay in finding a den. Newborn pups are small and helpless, so the wolves must dig an underground den at the base of a large tree to keep them safe and warm. Later the pups will cautiously begin to venture out, but for two months the wolf family's activity is focused on this one site. If you can find it then you have a good chance of filming wild wolf behaviour.

Luckily, I knew some people who could help. Prior to the start of filming I had spent several days with Ian and Karen McAllister, who have

In the high mountains (above): mother grizzlies with cubs spend a lot of time at high altitudes in the spring. These places are often rocky, snowy and devoid of anything for a bear to eat. It is thought that keeping to these inhospitable surroundings helps mothers and their young to avoid hungry males on the prowl, who will sometimes kill cubs for food.

been travelling and living on the coast, photographing and protecting the local environment, for the last 20 years. Ian knew the location of several different den sites in the forests and had some pretty good ideas where another half a dozen might be found, so I felt our chances were about as good as we could make them. The plan was to find an active den, move a camera hide into position, and then spend as much time as we could filming the wolf family as they cared for the pups when they first emerged.

Armed with our maps and notes from Ian, we headed off up the coast in our 21m (70ft) chartered boat. For four weeks we travelled hundreds of kilometers up and down the coast looking for wolves and their dens. We were out of luck. Every bay, inlet, or island that had had wolves in it last year didn't have them this year. After four weeks we had managed to see only three wolves on the entire coast and not even a hint of a den site. Earlier in the year I had installed two infrared cameras in dens that Ian knew about as it's not uncommon for a female wolf to use the same den year after year. However, if the wolf pack gets a new breeding female she is unlikely to use the same den as her predecessor. Whatever happened in those packs, no wolves used either of the old den sites that spring.

I knew that wolves are always tricky animals to predict. Year by year their behaviour can fluctuate dramatically. They will move to a totally different part of their territory and you can't find them. A pack may have a decline in a food source and some of its members starve. Other packs may raid and kill them. They may move into another pack's territory and take it over. They may merge with another pack so that a new hierarchy is established with different wolves that move the pack to a new area. Wolf society is constantly in flux.

Once again I had to admit defeat. This was turning out to be a not very auspicious start to the project. But I had one more chance. By the beginning of summer the wolf packs move their pups out of the forest and into the open bays and grassy estuaries. That was when I had my best chance to catch up with them, but that wasn't for another month or two. It was time to turn our attention back to grizzly bears.

Grizzlies in the grass

The late spring is a great time of year to film grizzly bears on the coast. The salmon are still hundreds of kilometers offshore, so the only things the bears can eat are plants. Bears are basically carnivores that have evolved to be able to live on plants. They don't have the digestive system of a herbivore, however, so they have to be very selective about what plants they eat. In June, the place for them to find the perfect food is on the grassy

Coastal wolf pups (left): wolves give birth in a den usually dug under a large ancient tree in the heart of the forest in the early spring. By June the pups are spending more and more time out of the den exploring the world around them.

estuaries. They come for a particular plant, a sedge that grows only in this intertidal zone. It's high in nutrients and easily digestible.

All of the big river estuaries along the coast will support their own populations of grizzlies. The trick is finding bears that haven't been hunted by man and learned to fear us. In order to get the shots that we needed, we had to find bears that weren't going to run at the first scent of us. It's almost impossible to keep yourself hidden from a bear for very long. Their sense of smell is so acute that they will always eventually learn that you are there. I have found that the best way to approach bears is to be completely open and visible. I don't sneak around and try to hide from them. I let them know I am there and carefully and slowly allow them to adjust to my presence and accept me. This lets me film natural behaviour and allows the bears to get on with their day without interruption.

Brown bear
Ursus arctos

A number of species of the brown bear can be found around the world, though populations outside North America are generally small. In North America brown bears living in the interior of the continent or in the Arctic often known as the "grizzly", because of the light-tipped guard hairs that give it a "grizzled" look.

Grizzly bears have humps of thick muscle on their backs and slightly curved claws. Both adaptations enable powerful digging, which is necessary to unearth roots and tubers, during the leaner months of April and May. The protein-rich sedge grass that is a staple part of their diet in the summer becomes more fibrous and less palatable just as the first of the autumn salmon begin to appear. The return of the salmon marks a time of abundance necessary for all bears to fatten up before winter sets in. Males travel vast distances in search of food, den sites, and mates, making the health of a grizzly population a good indicator of the health of the forest.

KEY FACTS

Size	up to 800kg (1750lb), though few are much more than half that.
Location	one in four grizzlies in North America lives in British Columbia.
Diet	omnivorous: roots, tubers, sedges and grasses; fruits and berries; salmon and nuts; also will predate on young and old ungulates like moose, deer, caribou, and elk.
Claws	can be 10–12.5cm (4–6in) long, slightly curved, and adapted for digging.

Surveying the scene (right): a young male grizzly bear inspects his coastal river home. Male cubs will eventually leave the area where they were raised, while female cubs often stay and share the mother's home territory.

A long childhood (above): a grizzly cub will spend the first three years of its life with its mother, following her and learning everything it needs to survive a life in the wild.

Fast runners (above): over short distances, grizzly bears have been clocked at speeds of up to 50km/h (30mph).

Bears love the water (above): coastal bears will spend most of their life in and around rivers. In the hot summer months they often laze around and play in the water to cool off.

Mothers and cubs

June is also the breeding season and males are on the prowl for receptive females. Mother grizzlies keep their cubs with them for three years, so they reproduce only once every four years. This is a stressful time of the year for mothers with cubs. They have to avoid the big males who are searching for females who are ready to breed. We travelled to several different estuaries and filmed male and female bears together. We also found mothers with two- and three-year-old cubs, but this was not what we were really after. We wanted to film mothers with tiny cubs, as they are the most vulnerable members of the bear population, the ones whose survival is most closely linked to the return of the salmon. Mother bears nursing young cubs put lots of energy into milk production and, after living on a subsistence diet of plants, they are badly in need of the salmon when they arrive. Also, the young cubs have to eat lots of fish if they are to put on enough weight to survive their first winter's hibernation.

The big problem that mothers with new cubs face in June is the congregation of so many bears in the estuaries. The cubs would be vulnerable to predation by an aggressive male, so the mothers tend to keep their youngsters hidden in the forest during this busy time. Again we searched up and down the coast for mothers with new cubs, but once more we struck out. It was terribly frustrating. Despite months of work and thousands of dollars spent, I had very little footage to show for it. I decided to call the spring shoot off early and bank some of my filming time for later in the summer when the salmon returned.

Familiarization (above): it's best to let the bears know you are there and allow them to come as close to you as they feel comfortable doing.

Just for fun (right): young grizzly bear siblings love to play. On the grassy estuaries in the spring I have even seen unrelated four and five-year-old bears meeting up and, after a few days of getting used to each other, starting to wrestle and play together.

Rendezvous wolves

In the early summer I had an excited call from Ian McAllister, saying he had found a wolf pack in a small bay not too far from Bella Bella. He hadn't see them, but he'd heard them howling. There were a number of adults and some pups. It was a rendezvous site; a place where the adult wolves leave the pups with a babysitter, usually one of the younger adult wolves in the pack, while they go off and hunt. At the beginning of the summer, when the pups are big enough, the adults move them from the shelter of the den site and set up rendezvous sites on the edge of the forest along the shore.

A rendezvous site is also a location the wolves don't venture far from, so you can stake it out and know that if you wait long enough you're likely to be rewarded with wolf footage. It was exactly what we were looking for. This was still early in July and I wasn't due to return for a few more weeks, but it boded well for our chances of finding wolves to film this summer.

I wanted to have as many options for filming wolves at these sites as possible, so I asked Ian and Karen to take their boat and conduct a circuit of the central coast, checking out the status of every wolf pack they knew. Some areas would be much easier to film in than others. Spending a week or more hiding out on the edge of a wolf rendezvous site was going to be difficult. How the wolves use the area, where they travel into and out of the site, and where the pups tend to wait and play all affect where you can set up your hide. You have to be able to see the action without being spotted and get into and out of it each day without spooking the wolves.

I think Ian, Karen, and their two young children were happy for the excuse to take off on a trip up the coast. In 10 days they covered over 500 nautical miles as they checked out 20 different wolf territories. But when I arrived in Bella Bella the day after they got back, the news wasn't good. They had not found a single wolf pack. Not one. In the past nine years Ian had never known a summer without locating at least one wolf rendezvous site. Our only hope was the pack that he had heard at the beginning of July, but it had been almost two weeks since he last checked on them.

We headed over to the site immediately in Ian's boat. It was early evening, the perfect time for the wolves to be up and about after a day of napping. Although not strictly nocturnal, wolves are more active at night and more likely to sleep during the day. We drifted quietly into the bay and waited. We didn't see or hear anything. Finally we tried a howl. Wolves will often respond to a well-done human imitation of their call. Ian tilted his head back and let loose a long, well-modulated, and well-toned howl. We sat back and waited. Nothing. He tried again and this time I chimed in a howl of my own. Sometimes two howls can get them going. Again

Early morning near Bella Bella (previous page): this magical stretch of coast has hundreds of inlets, any one of which could be hiding a pack of coastal wolves.

Rendezvous site (left): a family of coastal wolves brings its pups out of the forest in early summer. The pups are left here on their own or with a babysitter while the adults hunt.

we waited, silently. The air was still, so the howls would carry for at least 2–3km (1–½ miles). But there was no response. Not a good sign. It didn't mean for certain the wolves weren't there, but it wasn't looking promising.

After that Ian and I spent a few hours plotting out a course that would take us into every possible wolf territory that he hadn't already checked out. We spent the next couple of days checking, but with no success. We jumped on every lead we heard about from friends, colleagues, and fishermen working in the area. Every place we went, we found neither hide nor hair of a wolf. There were no responses to our howls. No tracks. No scats. Nothing. It was as if all the wolves on the coast had suddenly disappeared.

We went back and again tried the bay where the wolves had been. I howled and howled, but got no response. I was certain the pack had moved somewhere else. A rendezvous site will often last only for a few weeks or less. If the adults make a big kill somewhere nearby they will often move the pups to the kill site rather than carry the food back to them. Something like this must have happened, because the wolves that had been here three weeks ago were gone now.

Coast wolf
Canis lupus ligoni

A unique sub-species of grey wolf lives on the coast of British Columbia. Unlike other populations of grey wolves, coast wolves get more than 75 per cent of their food from marine resources such as salmon, beached whales, and seals. They are about 20 per cent smaller than continental wolves, perhaps an adaptation to the thick foliage they have to move through or because their principal land-based prey, the Sitka black-tailed deer, is much smaller than the prey that wolves hunt elsewhere. The hair of coastal wolves also appears to be coarser and better at shedding water, perhaps to cope with the abundance of water (both fresh and saline) on the west coast.

These wolves have also developed behaviours which allow them to operate in a coastal environment. They range over vast territories and are sometimes, perhaps often, forced to swim across sizeable saltwater channels. Scientists even joke that the coast wolf should be classified as a marine mammal because they are so unphased by making the transition between land and sea.

KEY FACTS

Height about 50–75cm (20–30in) at the shoulder.
Weight about 20–50kg (44–110lb).
Lifestyle very social, living in packs of 12–15 individuals and mating for life.

Last chance to film...

I had one more river system to check out, but it was a two-day boat ride there and back. If we didn't find anything, it would be time to pull the plug on this trip. I kept my fingers crossed and tried to stay optimistic. When we finally got to this remote river system we spent most of a day looking around but found no sign of wolves. We turned the boat around and headed for Bella Bella. I had to hope that when I came back in late August for the salmon season my luck was going to turn around or this film was going to be a bust. Not something I could imagine explaining to my colleagues at the BBC, who had spend hugely on the project, believing in my ability to bring back some spectacular footage from this difficult location.

Although I had known from the start it was going to be tricky, things were turning out to be much tougher than I could have imagined. But my experience has taught me never to give up. Successful wildlife filmmaking relies on a healthy dose of optimism. The greatest moments are, as Dr Johnson famously quipped about a man's decision to marry for the second time, "the triumph of hope over experience". Sometimes, when it seems that you are not getting anywhere and failure upon failure is heaping one on top of another, you just have to persevere and try again. Sometimes that is when you manage to get that one-in-a-million shot.

On the last day, as we were heading in to Bella Bella to catch our flight home, I decided to try the little bay where Ian had heard the wolves weeks before, one last time. It was almost a perfunctory visit, but I felt I had to cover all the bases I could. We paddled into the bay and I tilted my head back and let loose a long, low howl. Almost immediately wolves howled back. One, two, three, and then the sound we were longing to hear – pups! We could hear their tiny, squeaking voices amongst the deep-throated adults. The pack was back. Cancel our flight home! We were staying!

It turned out to be an ideal location for filming. We quickly figured out where the wolves were spending most of their time and found a good hiding place where we were out of sight and yet could see most of the bay. Most of the wolf traffic seemed to be coming and going on the other side of the water from our hide location, so they were less likely stumble upon us.

For the next week we filmed the wolf pack. There were four pups – two black ones and two tan ones. One of the black pups had different coloured

Coastal wolf (left): different from other North American wolves, this subspecies derives more than 75 per cent of its diet from marine-based food such as salmon or seals. Coastal wolves also have shorter, coarser fur than wolves further inland.

eyes – one blue and the other a golden yellow – something I had never seen before. The pups usually had two babysitters with them and we got some lovely footage of them playing and chasing each other around the estuary.

Although we had both adults and pups come up to with 3m (10ft) of us, they never figured out we were there. Our hide was positioned beneath a large spruce tree, which must have largely masked our scent. We were also downwind from the prevailing wind direction. A couple of times when one of the babysitters got close, he must have sensed that something was watching him because he would suddenly start to bark-howl, a sign of distress, and look around nervously, but he never locked onto our location. Eventually when he didn't see or smell anything else he would settle down and everyone would go back to what they were doing and forget all about us. It is a wonderful and satisfying experience to watch and work around wild animals without disturbing them. It was also our first big break in the filming for this project and it set us up for the busy and exciting autumn salmon season that was just around the corner.

Spawning (above): female salmon dig a shallow nest called a redd in the gravel of the spawning stream. When a female begins to lay, the male salmon moves in to spread his sperm over the eggs.

Mussel bears and spawning salmon

By August the salmon were finally beginning to arrive in the inlets and bays along the coast. It's generally thought that the schools of salmon heading for the western coast of North America swim south off shore along the continental shelf. The various races of salmon, those of each species coming from a particular river or stream, peel off when they reach the latitudes where their home waters flow into the sea. They navigate the maze of waterways and inlets branching off into smaller and smaller groups as they near the waters of their birth, finally arriving at the mouth of the river from which flows the fresh water they were born in two, four, or more years ago. How they do this is again a mystery. Do they have some sort of internal compass that charts their way to sea and back again, or can they smell or somehow otherwise detect the particular flavour of their river system from all the thousands of others they pass along the way? Studies have shown that a salmon's sense of smell is extremely sensitive. They can detect one part of a chemical substance in 80 billion parts of water. That must be something like the equivalent of sitting in England and smelling fresh baked bread at my home in British Columbia.

Salmon, however, are not flawless at finding their way home – they do sometimes seem to make a wrong turn. I have seen river systems where only 1000 pink salmon returned to spawn. Then two years later, when I was expecting about 1000 or so to show up, 50,000 appeared. Where did all the extra fish come from? And the reverse can happen as well: runs where tens of thousands are expected and only hundreds return. Also, I have seen thousands of pink salmon trying to get into creeks and streams with waterfalls so high that no salmon could get over them. Obviously these fish were not born in these creeks, as, like them, their parents could never have made it over the falls to spawn. It's not a perfect system, but that it can work at all is still a miracle to me.

Once the salmon have made their epic salt-water migration – avoided all the sharks, killer whales, and seals that have been trying to eat them; dodged the miles and miles of nets, lines, and hooks that man has out there to catch them – they reach the fresh water and can start their journey upriver. This is the part of the salmon's journey that we are most familiar with, the valiant struggle against the odds as they strive to reach their spawning grounds and lay their eggs before they die, all the while avoiding the claws of the bears, eagles, wolves, and other creatures out to get them.

The salmon's journey

Our filming in August began at the start of the salmon's journey into fresh water, the point where the river meets the sea. In some river systems, the biggest barrier to the salmon is right at the start – a waterfall at the mouth of the river. We travelled north of the village of Hartley Bay where the salmon were leaping over a 10m (33ft) waterfall right at sea level. Normally that would not be possible: it's just too high. But here the salmon have two advantages – first, they are fresh from the sea and filled with energy. Second, because the falls are right at sea level their height varied according to the rise and fall of the tide. The tides are big in this part of

One of the salmon's main ocean predators (above): salmon sharks (*Lamma ditropis*) grow up to 3.6m (12ft) in length and are one of the fastest fish in the sea, reaching speeds of 50 knots.

the world – 7m (23ft) or more – so at high tide the 10m (33ft) waterfall was shortened to 3m (10ft).

We watched these fresh silver-coloured fish shooting out of the ocean at the base of the falls as if they were being fired from a rocket. These were powerful sockeye and coho salmon – 4–5kg (9–11lb) bullets – blasting 3–4m (10–13ft) out of the ocean, arcing to land at the top of the falls. One of the biggest problems that salmon have in getting over falls is all the air mixed in with the water, which is what gives a waterfall or rapid that white foamy look. If the salmon land in the white foamy stuff, there isn't enough resistance for them as they push their powerful tails against the water to propel themselves forward. They need the clear dark water to have any hope of making it. Here there were sections of clear, fast-flowing water near the lip of the falls. If the salmon could land in that, they could power themselves up and over the top. Their strength was unbelievable. The water was so fast that there was no way a human, or a bear for that matter, could hope even to stand up in it. Yet the salmon could hold themselves in it and with an extra burst of energy push themselves through and over the top. Awesome.

Not only do the salmon have to cope with the physical barrier of the falls and rapids, they also have to avoid the snatching mouths of predators. If you are a hungry bear, these are great places to fish. Bears come from miles around to wait for salmon as they struggle to make it over these barriers. Throughout the coastal forest in British Columbia there are many falls where it's possible to film bears catching salmon, but the most famous are in Alaska – places like Brooks Falls in Katmai National Park and the McNeil River. At Brooks the bears can stand right at the top of the falls and literally pluck the salmon out of the air as they leap.

Once the salmon arrive at their spawning grounds, the females find a patch of river bottom to lay their eggs. Studies have shown that some salmon will return to exactly the same patch of river gravel in which they were born. The spawning areas are usually defined by the substrate of the river. For most successful protection and development of the eggs over winter, a bed of gravel, not too big and not too small, is required. Substrate that is too fine and silty will cover over the eggs and smother them; too big and there's not enough protection from the over-winter flow of the river and the eggs will be washed out to sea.

Females stake out a patch of gravel on the river bottom and keep all other females away. Turning on their sides, they use their tails to dig out a shallow depression in the gravel, called a redd. Males move into position

Salmon shark
Lamna ditropis

Salmon sharks are part of the same Lamnidae family as the great white shark, and although not as big as their infamous cousins they show many of the same characteristics: fairly large, powerful, warm-bodied (endothermic), and streamlined predators adapted for high-speed swimming. Reports from the US Navy have said that salmon sharks can exceed 50 knots, which would make them one of the fastest fish in the ocean.

Salmon sharks inhabit coastal and oceanic waters of the northern Pacific Ocean. Their annual migration, which takes them from the coast of California to the Bering Sea and back again, is thought to be linked to prey abundance, temperature, and suitability of environment for breeding. However, some salmon sharks inhabit Gulf of Alaska waters during all months of the year.

Despite their name, these sharks are opportunistic predators that feed on a wide variety of prey depending on availability throughout the year. In addition to fish such as salmon, herring, capelin, and sablefish. They have been seen to snatch sea otters and marine birds, but this is rare.

KEY FACTS

Length about 2m (6ft 6in)
Weight about 200kg (440lb)
Body heat unlike most fish, the salmon shark is endothermic – it can generate its own heat – and is able to maintain its body heat at about 7–11°C (13–20°F) above that of the surrounding water.

Fishing made easy (left): grizzly bears at Brooks Falls in Alaska are able to stand on top of the falls and the salmon will literally jump into their mouths.

behind a female sitting over her redd and wait. There is much jostling for position, with males battling each other for this prized spot. They have to be ready to move in and spread their sperm over her eggs as soon as she begins to lay them in her gravel nest. When the moment finally arrives, eggs and sperm mingle together in a white cloud as the now fertilized eggs settle into the depression in the river bottom. If the female has done her job properly, the eggs will settle into the nooks and crannies between the gravel, where they will spend the next six months. Then the following

Ideal spawning habitat (left): good conditions are defined by the flow of water and the size and shape of the river substrate, where the salmon's eggs will be oxygenated by an adequate flow of water and protected in the nooks and crannies between the gravel.

spring the newborn salmon wiggle up through the cracks and gaps in the gravel and emerge into the flow of the river, ready to be swept downstream and out into the vast Pacific Ocean to begin the cycle all over again.

Why the salmon die after spawning is not precisely known. Some salmon travel more than 1000km (650 miles) upstream to return to their natal area. In so doing they expend huge amounts of energy, and by the time they reach their spawning areas they are usually physically very battered. The depletion of their energy reserves is so complete that they don't have enough left to make it back downstream to the ocean. They die from sheer exhaustion. Some scientists believe that salmon have evolved to die in order to supply nutrients from their decaying bodies to the next generation.

I walk up the forest stream overflowing with spawning salmon so thick that it seems possible to cross without getting your feet wet. Gulls, eagles, ravens, crows, jays, and a dozen other species fill the air with their cries. And surrounding it all is a forest of giant trees harbouring more biomass than any other habitat on earth. Here nature is showing her extravagance.

Waiting for rain (above): pink salmon school in large pools, waiting to continue their journey upstream to their spawning grounds. Low water often makes the journey difficult, if not impossible. When the rain comes it will bring the water level up.

Feeding time (above): Grizzly bears are the main predators of salmon in the Great Forest. A single adult will consume over 1000kg (2200lb) of salmon during the spawning season.

At the peak of its run, a salmon stream is so overfilled with life that it cannot contain it all and it spills out along its shores. At this point there are almost equal numbers of living and dead salmon. Sometimes the shores are so deep with washed-up bodies that you can't see the ground for hundreds of metres. It's a seething mass of maggot-infested, decaying tissue. And yet everywhere there is life. Entire insect communities, unique to the coastal forests, have evolved to take advantage of this abundance. I stop at one point and count a dozen grizzly bears feeding on the living and dead salmon in front of me. Further upstream I can hear wolves howling. Spending time in the temperate rainforest during the salmon run is to experience a side of nature that few people get the chance to see.

When the salmon finally arrive in the rivers, the bears go into a feeding frenzy. This is what they have been waiting for since emerging from their den. Bears will eat and eat and eat salmon, barely even stopping for sleep. They have a unique ability to turn off their feeling of being full. They can consume literally hundreds of kilograms of salmon a day and never feel sated. It's very helpful to them when they need to put on weight quickly, which is crucially important to their ability to survive the winter. If they don't have enough fat reserves when they go into hibernation, they will die.

Salmon predators: bald eagles (*Haliaeetus leucocephalus*, left above) and coastal wolves (left below) feed extensively on salmon during the spawning season. They help distribute marine nutrients from the salmon's body throughout the forest.

Food in the mud (next page): a mother grizzly bear leads her young cubs over a tidal estuary in late spring. Here there may be clams to be dug from the mud, sedge or roots to be taken from the banks or berries gathered from the bushes.

A glut for the grizzlies

When we began our filming in early September, there were 20 different grizzlies in the river system and we got some wonderful footage of mothers and cubs as they went about chasing and catching salmon. We focused on one mother who had two new cubs and another older mother with a three-year-old cub almost as big as she was. The two young cubs were wonderful to watch. They would follow their mother, sampling every fish she pulled out of the river. She would calmly leave it for them and go and get another. When they eventually noticed she had another salmon they would drop the one they were eating and rush over to try the new one. They got bored eating long before their mother did, and then they would chase each other around, wrestling and tumbling. After hours of feeding the mother finally lead the cubs back into the security of the forest to rest. But after only a two- or three-hour nap she would be back out there, pulling salmon out of the river.

The mother with the three-year-old cub was the dominant matriarch of the river, displacing other bears from the best fishing sites whenever she showed up. She was very calm and unconcerned by our presence. She would walk past within 10m (33ft) of our filming position and didn't mind if I moved around her to get a different angle. Her cub was very inquisitive and would come even closer, stretching his neck forward and sniffing us, curious as to what we were all about. But he was very respectful and didn't come near enough to disturb us or our equipment. It was with these two that I finally managed to fulfil a small ambition of filming grizzly bears climbing trees after apples. There is a species of wild Pacific crab apple that grows along the coast and is a favoured food of bears in the autumn. I had often seen these small trees with broken limbs from the effects of bears feeding, but had never been able to film a bear actually in a tree.

Arriving on the river early one autumn morning we couldn't see any bears in the water. Then as we looked across the estuary we noticed

Black bear (above): more numerous than grizzlies throughout the Great Forest, black bears also rely on salmon as a mainstay in their diet. With their shorter, hooked claws, they are more comfortable in trees than the grizzlies, whose longer claws are adapted for digging.

something big and brown high up in the apple trees. It was the three-year-old cub. We raced over and, sure enough, he was clambering about in the top of the tree, about 8m (25ft) off the ground. The crab apples were small but numerous and the bear was contentedly filling his belly with this tart but juicy fruit. Below him his mother was standing on her hind legs, reaching up to pull down the lower branches to eat her fill. It must have been a nice variation in their diet. I have noticed that bears will often switch over to berries or other shrub fruit during the salmon season. I guess they can get tired of the steady supply of fish.

September was turning out to be a dream month: I was finally filling the tapes with footage of bears, salmons, eagles, and wolves as they took advantage of the abundance of salmon. But there was still a lot to do.

Black bear
Ursus americanus

Black bears are the most widely distributed of British Columbia's large mammals. Virtually the entire province, including the outer coast and islands, is occupied black bear habitat. Humans have settled 8 per cent of the province, but even parts of the densely settled areas still support black bears.

Despite their name, black bears are not always black, though it is the most common colour. There is a white sub-species, known as the spirit bear, and another variation, one of the rarest, has a pale-blue coat and is known as the glacier bear. Like grizzlies, black bears take advantage of the salmon run to feed themselves up before winter kicks in.

KEY FACTS

Location throughout North America from northern Mexico to all the provinces and territories of Canada except Prince Edward Island.

Weight males normally around 115kg (250lb), though can be as much as 275kg (600lb), females about half the size of males.

Diet omnivorous, feeding on insects, nuts, berries, the young of deer and moose, and, in British Columbia, the salmon run.

One of the challenges of filming wildlife today is to capture behaviours that have been seen and filmed before in new ways. With better cameras and specialized equipment we can see into the natural world in ways we couldn't a few years ago. I had filmed grizzly bears catching salmon many times. What I wanted to do for this film was to show it from the point of view of the salmon. I wanted a shot that showed the salmon under the water at the same time as you saw the bear charging towards it above the water. In order to do this I needed a small underwater camera that I could hide in the stream and control remotely from a distance. It was going to be a wide-angle shot in the middle of the action, set half in the water, half out. When the bear started chasing after salmon we would turn the camera on. It was – we hoped – going to be the ultimate grizzly-bear-fishing shot.

As with a lot of technology, it didn't work perfectly first time. I had an ideal spot picked for the camera, where we had seen lots of fishing activity and the water levels were just right for the shot. The problem was the cable. It had to be long enough that we could be sufficiently far from the fishing site not to bother the bears, but not so long that the video signal travelling along it would lose quality. It was a fine-tuned thing and any kinks in the wire would make a huge difference to how well we received a signal.

On the day of our big shoot I had stretched out all our cable – about 100m (330ft) of it – along the bottom of the river bed. I knew how curious bears can be and made sure that the cable wasn't too obvious. But I guess I wasn't careful enough. Moments later, even before I had a chance to get back to the computer and set up the shot, the mother and cub came up the river towards us. Both bears were walking in the water. They walked right past where my assistants were setting up the computer, coming towards me in the river. The mother bear was looking for fish and since there were lots of dead ones around she was busy picking out ones that appealed to her.

The cub, on the other hand, seemed to be quite interested in what I was doing. I could see him looking intently at me as he approached. Towards the end of the salmon season when the bears are getting lots to eat they often become much more playful. It's as if the pressure of frantic feeding that grips them early in the season has worn off as they realize that they are now going to be fat enough to survive their winter's hibernation. The cub wasn't hungry. He was looking for something new to do. He could see that I was doing something in the water and whether this prompted him to look down into the river bed I don't know, but he suddenly spotted the black cable. Before I had a chance to react he reached down and picked it up in his mouth. I knew what was coming next. The first thing a bear will do with anything new he finds is test it to see if it might be good to eat.

Fish farming

Salmon farm (above): these are located by the north-east and west coasts of Vancouver Island. British Columbia is the fourth largest producer of farmed salmon in the world.

Also known as aquaculture, this is the aquatic form of agriculture in which stocks are cared for, raised to marketable size, and then harvested for processing, sale, and consumption. The industry's argument for aquaculture is that it creates employment and economic growth. With the decline of wild stocks and commercial fishing, aquaculture has become a mainstay on Canada's east and west coasts.

Highly controversial, salmon farming has gone on for a number of decades all along the north-west coast of North America, and as a result, has polarized communities that have traditionally relied on salmon, as well as those that have more recently become dependent on the aquaculture industry. With such a huge part of North America's salmon fishery balanced on a knife edge, it's no wonder that heated debates continue.

Despite the industry's economic importance, there are several main issues arising from salmon farming. The first is that a number of companies rely on growing and harvesting Atlantic salmon in Pacific waters. This has raised grave concerns among scientists and environmentalists, who claim that as many as 1 million farmed fish have escaped from their pens since the 1980s, endangering the wild populations by introducing rogue genes into their gene pools. Another major problem is that wild salmon may be being driven to extinction by parasites that infect nearby farm salmon stocks. On the other hand, there is no denying that the fish-farming industry provides employment in an area that needs it. There are no easy answers to questions like this. Recent scientific review of salmon aquaculture studies has concluded there isn't enough evidence one way or the other to say that salmon farms are negatively affecting wild salmon stocks. More time and money needs to be spent on the question, but in the meantime we should err on the side of caution when it comes to licensing fish farms in areas that still have abundant wild salmon populations.

A lucrative business (above): in 2004, the British Columbia salmon aquaculture sector produced more than 61,000 tonnes of Atlantic, chinook, and coho salmon.

I leapt up and started waving my arms at him and yelling at him to drop it. I stumbled down the stream towards him, but I was too late. Chomp! He bit down on the cable just as I startled him, then dropped it into the stream, and ran to catch up with his mom. That was the end of getting my ultimate grizzly-bear fishing shot for this year.

Ghost bears

As September began to wind down, the salmon run was nearing its end. Most of the fish had spawned by now and the bears had switched from fishing to scavenging carcasses from the river bottom. The autumn storms were beginning in earnest and the river levels were rising, washing the remaining carcasses back to the sea. It was time to call it quits for the year.

As our boat slowly chugged down the long inlets on our way back to Bella Bella for our flight home, I found myself wondering how much longer we would be able to come to places like this and see the spectacle of the salmon run. I had first filmed on this coast 20 years earlier and as I watched the fresh logging cutblocks appear around the bend in the inlet it wasn't hard to visualize how much had altered. Many of the river systems that I went to film in still had salmon and bears much as I remembered them, but others had changed – and not for the better. New logging was continuing to make inroads into pristine river systems. Salmon numbers were down from what they were even 20 years ago. In addition, the increasing popularity of eco-tourism has greatly increased the presence of tourists along much of the coast. It's hard to find the secluded and secret places that I remember from my earlier years on the coast.

During the autumn, my wife Sue and our children Chelsea and Logan all came out to help with the filming. Some 16 years previously Sue and I had lived for two years in tents on an uninhabited island on this coast, making a film about the spirit bears or ghost bears, a rare white sub-species of the American black bears that lived there. Chelsea was just six months

After spawning (right): chum salmon lie washed up in the estuary. Eagles, ravens, gulls, bears, and wolves, as well as flies and beetles, will scavenge their bodies. The remains will be washed out into the stream and may even provide nourishment for their growing eggs.

old when we went to live on the island, where she learned to walk and talk and where we had many wonderful experiences with the white bears. One day on the last filming trip of the autumn we were all together trying to film salmon on a small stream in the heart of the forest. It was dark and dripping. It had been raining most of the day and we were tired and wet. Suddenly I looked up and saw a ghost – ambling along the shore towards us was a white bear. It seemed to glow in the gloomy forest light as it hunted along the shore. It had been 16 years since Sue and I had last seen

Spirit bear (left): this rare white colour phase of the black bear can be found in certain areas along the British Columbia coast. It had been 16 years since I last saw one.

one. Chelsea had been a year old then and couldn't remember; Logan, now aged 13, had never seen one. We stood transfixed and silent as the bear made its way slowly towards us, head down, sniffing for carcasses washed up along the shore. After only two or three minutes it turned off into the woods and disappeared.

The most productive landscape on Earth is still here; living, growing, dying, and being reborn, much as it has since the last ice age. Sixteen years on, despite the relentless pace of logging and over-fishing, the Spirit Bear can be seen hunting along the forest streams. I hadn't been sure if my children would ever get the chance to see such a wonderful and magical creature as this. There must be something right in the world if it is still possible. If we can learn from our past mistakes and moderate our appetite for trees and fish, this unique ecosystem will remain healthy enough for thesalmon to continue to fight their way up the rivers of this ancient forest. Who knows, maybe one day I'll even be able to show a Spirit Bear to my grandchildren.

A special habitat (above): a grizzly bear of the Great Forest has the Pacific salmon to thank for making his home some of the most productive land in the world. We need to ensure that our appetite for salmon and timber do not compromise it.

The Great Tide
South Africa

Africa is the continent of great animal migrations: at different times of the year wildebeest, zebra, and even elephants migrate over hundreds or thousands of kilometres searching for watering holes. But in South Africa, the greatest animal “migration” on the planet takes place in the sea, starting near the Cape and ending some 2000km (1250 miles) to the north, near the border with Mozambique. This ocean-borne phenomenon is called the Sardine Run.

At the beginning of the southern winter, billions of silvery fish (the Pacific pilchard, *Sardinops sagax*) hellbent on following a burgeoning food source northwards along the KwaZulu-Natal coast, gather into huge shoals, any one of which may reach 20km (12 miles) long and 5km (3 miles) wide. These shoals attract the great predators of the deep, which congregate in an astounding underwater frenzy. Thousands of sharks, dolphins, fur seals, and Bryde's whales assemble for an orgy of feasting, while tens of thousands of Cape gannets swoop down from the skies. It is the greatest aggregation of predators on the planet. This is a marine predation spectacle, rivalling the drama that plays out with the wildebeest migration on the Serengeti plains in East Africa.

South Africa
Africa

South Africa is recognized as possessing some of the world's most important biodiversity ecosystems. The country is ranked the third most biologically diverse in the world, and its marine environment is no different. The two oceanic systems that dominate the Southern Cape (the Benguela and the Agulhas currents) give rise to a rich and varied marine life, with over 11,100 species – 15 per cent of the world's coastal marine species – having been described so far.

The marine habitats found along South Africa's stormy coastline include mangrove forests, sea-grass beds, kelp forests, and stunning coral reefs. Over 2150 species of fish are recognized. This ocean realm exists thanks to the unique conditions that occur where the Benguela and the Agulhas meet.

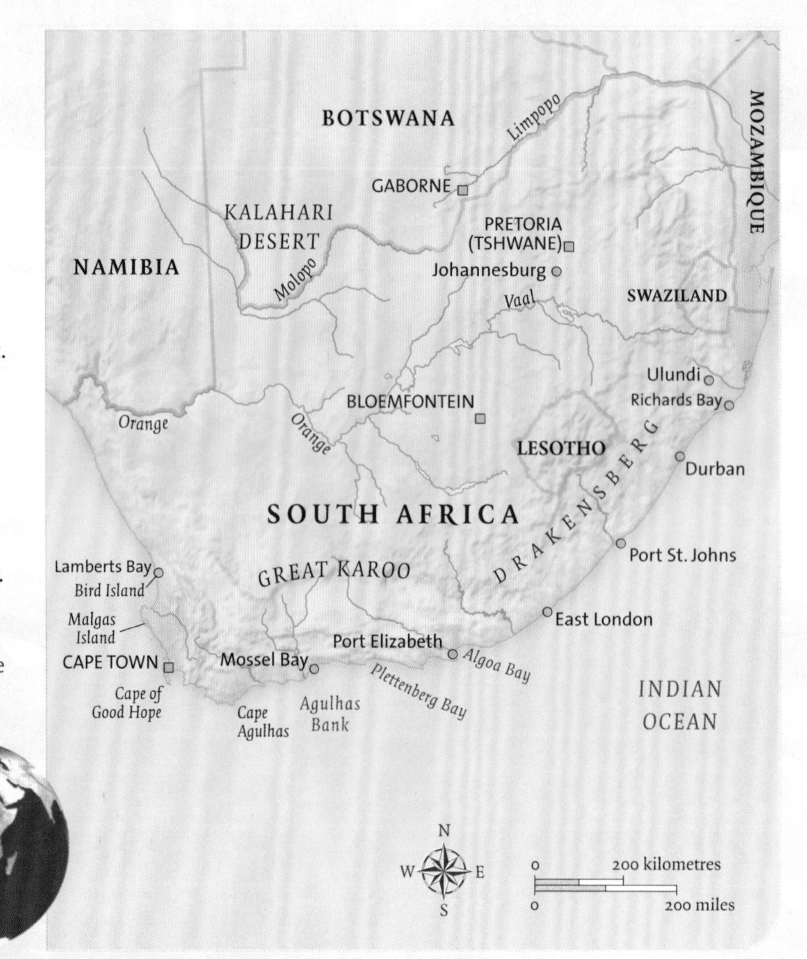

The Agulhas Bank, off South Africa (right): the place where it all begins. This is where the Agulhas and Benguela currents meet and is an important breeding ground for the sardines.

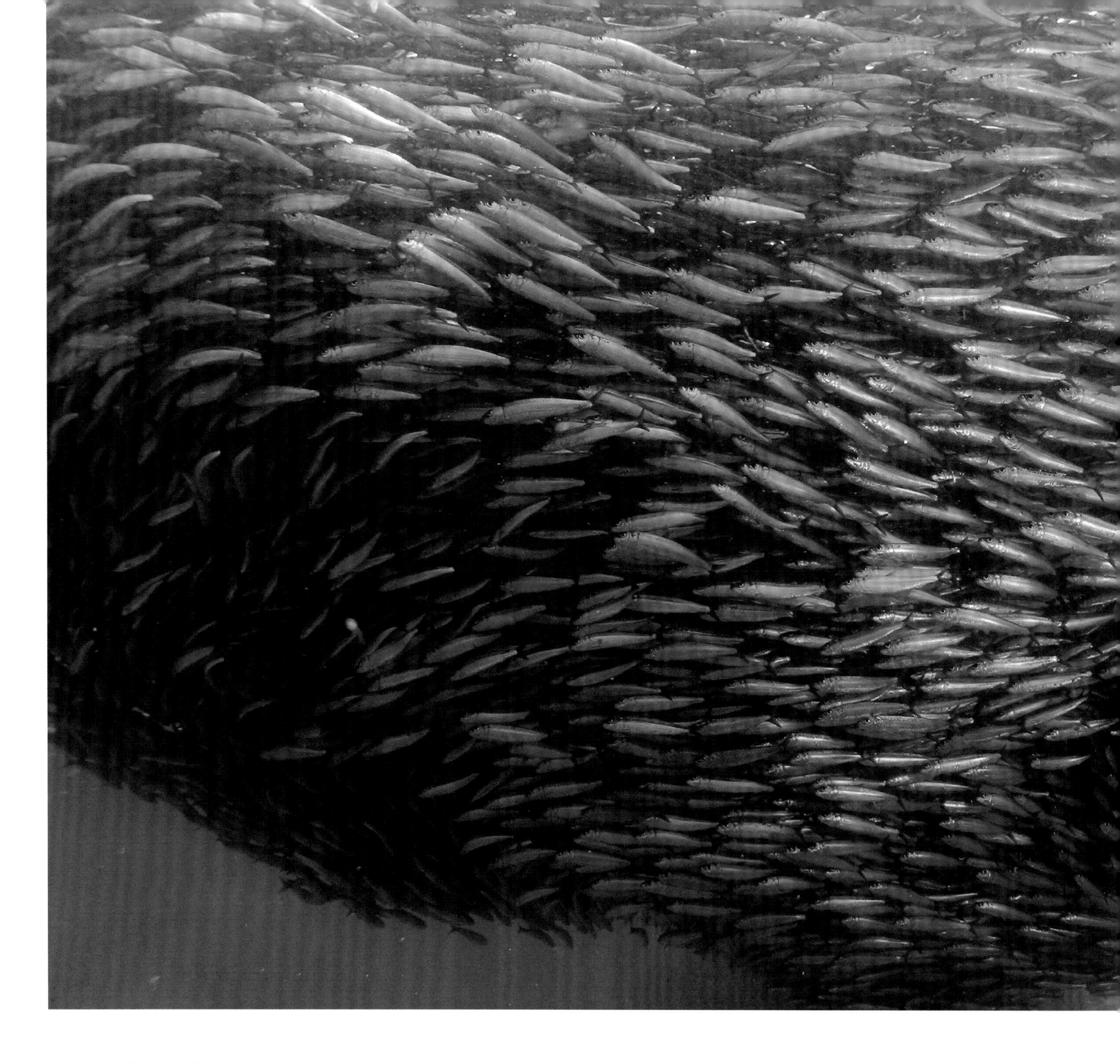

An unequal struggle

The sardines tenaciously follow their route north; their predators, who will inevitably win each and every battle, continue to harry them into the shallow waters of the Natal beaches, where humans pick up their share of the plunder. From the decks of tiny boats, fishermen cast their enormous nets over the water. This traditional form of fishing attracts the entire local population and the beaches turn into a gigantic, excitable jamboree.

But the sardine run is more than just a tragedy for the fish. It is the biggest marine spectacle on the planet, the greatest gathering of ocean predators. The lives of many thousands of dolphins, sharks, whales, seals, and sundry species of birds all depend on this seemingly suicidal sardine dash.

Two great ocean currents exert a powerful effect on southern Africa: the Benguela, which flows up from the Antarctic, and the Agulhas, which

A sardine bait ball (above): it may consist of hundreds of thousands of fish very closely huddled together. Balling up like this is the best way for these tiny fish to protect themselves against predators. The technique is thought to have evolved because it can confuse predators whilst hiding individuals within the swirling mass.

comes down from the equator along the eastern seaboard of the continent. It's the battle between these two mighty currents that dictates when – and if – the sardines run.

The Benguela first skirts the Cape coast before running into the Atlantic, west of Namibia and Angola. Its waters, which are cold but nutrient-rich, support a wealth of sea life. Conversely, the Agulhas current rises in the Indian Ocean and has all the characteristics of a river as it

Oceanography

Southern Africa's Agulhas Bank is the meeting place of two mighty oceanic currents: cold waters of the Atlantic Ocean are borne northwards along the west of the continent by the Benguela Current, and the warmer Agulhas Current of the Indian Ocean brings water southwards from the equator along the eastern side.

The Benguela forms the eastern boundary of the South Atlantic sub-tropical gyre, a circular ocean current flowing anti-clockwise between South America and Africa. This gyre sends water from Antarctica towards Africa's southern tip where it's forced northwards along Namibia's coastline towards Angola. As the water travels along this path, its upper layers are deflected by the earth's rotation, driving them offshore to the west, replacing inshore surface waters with colder, nutrient-rich water. This creates an explosion of life.

The waters of the Agulhas Current are also largely derived from a circular ocean current, in this case, the southwest Indian Ocean sub-gyre. Originating off Mozambique's southern coast, the Agulhas extends to a depth of 2,500m (8,200ft) and its main stream can be 60-100km (37-62 miles) wide. Transporting warm, tropical waters from the Indian Ocean southwards towards Africa's southern Cape, it can travel at a rate of 2.6m (8.5ft) per second, making it one of the fastest currents in the world.

Where these two ocean giants meet, cool waters from the Atlantic combine with warm waters from the tropics, resulting in one of the richest and most diverse marine ecosystems on the planet.

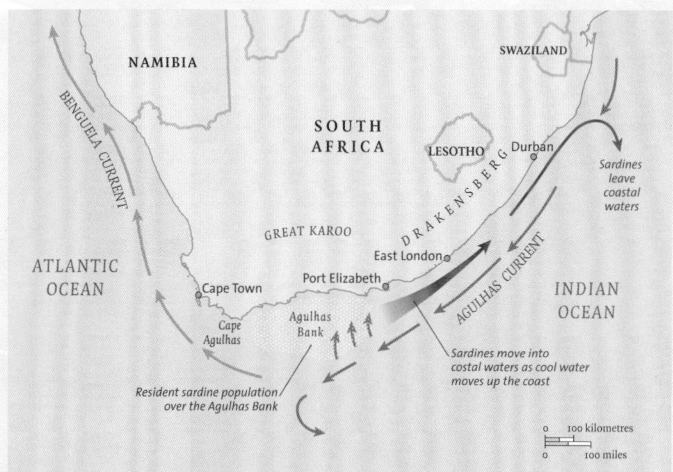

The currents (above): in winter the Agulhas current is pushed off-shore, allowing a tongue of cooler water to stretch up the coast – and allowing sardines to venture northwards.

rushes down towards the Cape of Good Hope. Along the coast, the high temperatures of the Agulhas support a great variety of corals, but further out to sea its clear blue waters can sustain little in the way of life – for as far as the sea is concerned, blue is the colour of the desert.

Where the two currents meet, near the Cape, the Agulhas raises the temperature of the sea by several degrees. And with the nutrient-rich waters of the south combining with the great diversity and milder temperature of the waters from the north, one of the richest ecosystems of our planet emerges. The temperature is ideal for plankton, the first link in the food chain and the vital foodstuff for the sardines.

The beginning of the southern winter reduces the strength of the Agulhas Current and causes it to lose momentum. Its temperatures drop and the Cape region no longer gets the benefit of the warm waters that normally take the edge off the chill of the Benguela. A tongue of cold water creeps along the coast, flowing north-east, against the current.

An opportunistic portion of sardines, which inhabit the temperate waters of the south all year round, can now take advantage of this extended area of cooler water. They leave Cape Province in huge numbers, shoal after shoal surging into their new environment along the coastline.

The Cape of Good Hope

Its name is the Cape of Good Hope, but it is also called the Cape of Storms, the point where the Indian Ocean meets the Atlantic. Situated at a latitude of 34 degrees South, the Cape has always been feared by sailors. The wrecks of ships spanning the centuries show why. Even today, despite improved navigational aids, ships frequently run aground all along its shoreline.

This is where Africa ends. To the south, there is no more land, nothing to break the constant battering of the giant waves which spring up in the "roaring" forties and the "furious" fifties around the Antarctic.

The few offshore islands, constantly pounded by storms, are home to large colonies of birds and marine mammals. On rocky outcrops further out, such as Seal Island and Geyser Rock, huge populations of fur seals occupy every single square metre. Colonies of African penguins, whose survival depends on the rich food supply in the water, have settled on a few beaches, as well as on some of these tiny islands.

Further east, sand dunes replace the rocky coastline. In Algoa Bay, Bird Island is home to the biggest colony of Cape gannets in the world. Like the albatross, these birds are the lords of the sea. Designed for life at sea, they can spend weeks or even months away from land as they wheel above the ocean in their hunt for food. An occasional beat of their wings is enough to keep them just above the surface of the water as they follow the sardines.

Cape fur seals (left) playing near their breeding colony of Robberg Peninsula, South Africa. The name means "Seal Mountain" and some 5000 seals can be seen here during the breeding season, sharing their fishing grounds with the local population of kelp gulls.

The wild Transkei coast is the final rocky hazard the sardines must face as they journey north. At the edge of the Mkambati Nature Reserve the cliffs of Waterfall Bluff plunge into the sea, marking the start of the offshore continental shelf, which is at its narrowest at this point. This is where all the characters of the sardine run come together. It is the home straight before the sardines reach the beaches of Natal that mark the journey's end. A heavy swell will drive a fair number into the hand of man, whilst others will turn back south, carried along in the depths of the Agulhas Current.

The run

It is not strictly accurate to call the run a migration; the sardines do not migrate like the wildebeest of East Africa do, in search of new green grasses, or the humpback whales that need tropical waters in which to give birth. The sardines are really just extending their territory, following a seasonal glut of food. In winter the cool waters carry the plankton on which the sardines feed further north than at other times of the year; and the plankton multiply even more thanks to the upwellings that bring nutrients up from the depths.

But the sardines themselves are a food bonanza for a whole range of predators. Large slicks of oil appear on the surface of the sea, betraying the presence of the fish far below. The Cape gannets, which by now have deserted Bird Island, tirelessly patrol the skies. Hordes of copper shark follow in the sardines' wake, as they too prepare to undertake the great journey toward the warmer waters of the north. But the maestro of this riotous symphony is the common dolphin, the architect of the bait ball feeding frenzy. During the sardine run, the dolphins, which normally fish at great depths for their staple diet of squid, gather near the coast in compact pods of up to several thousand individuals. They fan out in line across the ocean to hunt sardines. Once they have located the glittering little fish, a whole group dives in order to separate off a part of the shoal, which they then force up to the surface. The sardines' only means of defence is to confuse their assailants by forming a tight bait ball, because the predators then don't know which fish to target. Copper sharks, which have been waiting for the signal to attack, swim up from the depths. They can now take advantage of the free meal by pouncing on the sardines

Robberg Peninsula from the air (above) looking east along the South African coast. Behind the peninsula lies Plettenberg Bay, where we filmed Bryde's whales for *Nature's Great Events*.

Stay in school
Schooling behaviour of sardines

Sardines, also known as pilchards (*Sardinops sagax*), are cool-water-loving fish belonging to the herring family. They feed on plankton in areas of intense upwelling, and spend most of their lives in dense schools, some of which can contain hundreds of thousands of individuals.

The term "shoal" is used to describe any group of fish that remain together for social reasons; there are a number of distinct advantages to communal living, such as increased reproductive and foraging success. It is also thought that fish in a shoal benefit from increased hydrodynamic efficiency, allowing them to conserve energy by swimming in the slipstream of others and thus reducing the drag effects of the water.

The term "school" is more specific. It denotes a closely knit group of fish belonging to the same species that swim in a highly synchronized manner. In a school, not only do all the fish swim at the same speed in the same direction, but a consistent distance is maintained between all individuals within the group. When predators threaten a shoal, the fish almost always form into a school, and this is exactly what the sardines do. By gathering together in a tightly packed shimmering ball of silver, the fish have their best chance of avoiding the predators' hungry mouths – and as the size of the school increases, the likelihood of an individual sardine being targeted is reduced. The sheer numbers and synchronized movement of a school can confuse predators, making it difficult for them to pick out an individual fish from the twisting, flashing mass and then grab it before it disappears back into the group. This is why some predators (such as common dolphins) work to separate a smaller ball of fish from the larger school, making it easier for them to hunt.

But how do individual fish move together as one huge school, with each individual precisely spaced within it? The fact that most schools disperse after dark suggests that vision plays an important part in the behaviour – the lack of light prevents fish from moving effectively together, which in turn would explain why sardines descend back into the depths at night, away from their predators. However, fish also have a lateral line along their body – a sense organ used to detect movement and vibration in the surrounding water – which enables individuals to respond quickly to the movements of others. By using sight and lateral lines in this way, all the members of a school can know exactly what the other fish around them are doing. Therefore all each one needs to do is copy the movements of its neighbour, and these movements will be transmitted through the entire group.

Common dolphins "herding" a bait ball towards the surface (below): the dolphins are efficient predators, fanning out to make sure their prey moves in the right direction.

trapped near the surface. And above the sardines, the sky turns dark as the thousands upon thousands of gannets that have been following the dolphins overhead orchestrate their attack and plummet down onto the bait ball.

Other predators also take advantage of these rich pickings. Near the coast, bottlenose dolphins congregate, waiting for the sardines to swim past, and, further offshore, Bryde's whales take their share as they lunge with open mouth through the bait balls.

As the migration progresses, the shoals of sardines expand. Along the Transkei coast, they make one last stop at Waterfall Bluff. Here, the narrow continental shelf acts like a funnel for the sardines which, once they have swum into it, have no way out other than via the beaches of Natal.

Bryde's whale
Balaenoptera brydei

This is the second smallest of the rorquals (a group that includes the blue, humpback, and minke whales) and has comparatively small flippers for its body size. Bryde's (pronounced "broo-dess") whales are not migratory, but do move between inshore and offshore waters to follow food. Although they retain the characteristic baleen plates used by other rorquals to sieve tiny food particles from the water, they feed almost exclusively on shoaling fish and squid, often exploiting the activities of other predators by engulfing bait balls corralled by species such as common dolphins.

There is often confusion between the sei and the Bryde's whales, as they are similar in both size and appearance. However, Bryde's whales are unique in having three longitudinal ridges on their heads; all other rorquals have just one.

KEY FACTS

Length an average of 12m (40ft).

Weight maximum 25,000kg (25 tons).

Distribution found in the Atlantic, Pacific, and Indian Oceans, largely in warm temperate and sub-tropical coastal waters where temperatures average about 20°C (68°F).

A Bryde's whale breaches the surface (left) startling the gannets and dolphins nearby. Though Bryde's whales feed alone for most of the year, they join other predators in following the sardine run, sometimes consuming an entire bait ball in one gulp.

A party of gannets "on the prowl" (next page): from this height they can plunge on a bait ball with great force.

Sardine quest
Didier Noirot

The first time I saw a sardine in South Africa it was lying beside the N2 road, between Durban and Port Shepstone. A strange place to find a fish... It was in July 1996 and I was the cameraman on board the *Calypso*, a member of Commander Jacques Cousteau's team. I was looking for the beach where the sardines had fixed a rendezvous with the fishermen. Clearly I had arrived too late and the catch had already happened somewhere else on one of the great sandy beaches strung out along the Natal coastline. Sardines caught by local fishermen were already for sale at the side of the motorway.

Yet there had been nothing unusual about 1996: the great gathering of the sardine run had taken place at the usual time of year and if I wasn't there, it was quite simply because I didn't know enough about it. It was only a few years later, as I was filming the run for the BBC, that I realized that even the best specialists couldn't predict with any certainly when these fish were going to arrive off Natal. I have now twice been involved in filming a single sequence at a given moment on a specific part of the coast, for the series *RoboShark* and *Blue Planet*. The episode on the sardine run in *Nature's Great Events* hoped to give a picture of the migration as a whole, taking the film crews from the Cape of Good Hope to the north-east of the country. This would be a far greater adventure.

Trawlers

It is late on a December afternoon. Justin Maguire, a freelance cameraman and I decide to put out to sea with John, a professional fisherman living in the port of Gansbay in the Cape Western Province. We leave Gansbay on board his trawler, the *Lady M*. The crew, made up of about twelve men, is getting ready to spend yet another night at sea. The sky is heavy, with a light wind driving low clouds towards the coast. As we leave the port, we pass a

Sardines at the surface (right): sardines have no escape once they are pushed to the surface. Below and all around in the water lurk dolphins and sharks; above there are gannets waiting to plunge.

few fur seals playing in the kelp as a dozen gulls squabble over the remains of a dead fish. The sea is calm.

Tonight John and his men will be going about 30 nautical miles out to sea to cast their seine net over the shoals of fish. This is where they will find the sardine "harvest", and it really is no exaggeration to talk about a harvest of fish. The sardines collect in massive shoals which cover hundreds of square kilometres.

As the trawler makes headway, more seals swim up alongside it and eventually form a compact escort. They know why John is here and they are going to take advantage of it. Once the seine net has been closed, they will find it easy to catch the fish near the surface.

Night falls, and as the *Lady M* arrives on the fishing ground she starts her search. Eyes riveted on the echosounder, John calculates the size and

Cape fur seal
Arctocephalus pusillus pusillus

The South African or Cape fur seal ranges from along the coast of Namibia down to the western and southern coastlines of South Africa, and has an estimated population of 1.5–2 million. It can be found up to 180km (over 100 miles) offshore and dives to a depth of 400m (1300ft) to catch its prey. In addition to their usual diet, some individuals have also been seen feeding on seabirds such as Cape gannets and African penguins. The gannet population on Malgas Island is in dramatic decline because Cape fur seals are thought to be responsible for as much as 80 per cent of fledgling mortality.

Commercial hunting of Cape fur seals began in the 1600s and still occurs in Namibia today. Hunting season lasts for four months each year, from August to November, and quotas can be as high as 60,000 pups and 7000 adult males. The Namibian government maintains that this is necessary to protect its fisheries, although there is little evidence that culling populations in this way is effective. Namibia also trades in fur seal pelts and penises (sold as aphrodisiacs in traditional Asian markets).

KEY FACTS

Length	males up to 2.2m (7ft), females 1.7m (5ft).
Weight	males average 200–360kg (440–800lb), females 120kg (260lb).
Food	pelagic schooling fish such as sardines and anchovies, and cephalopods such as squid.

Cape fur seals (left) can hold their breath underwater for over seven minutes, which enables them to capture squid and small fish along the sea bottom. They also make themselves unpopular with fishermen by scavenging from their nets.

Cameraman Didier Noirot filming sardines as they shoal all around him (above): the water is clear and the fish seem calm, but they know that a shark might appear any minute now.

position of the shoal of fish, which shows up as red bubbles on the screen. 100m (330ft), 80m (260ft) down... still too deep. It's too soon to cast the net. Later, when it is completely dark, the plankton will begin its migration up towards the surface and the sardine will follow into the shallow waters.

John puts the engine on full throttle. The aft winch spins out the net which opens like a fan. The helmsman jams the tiller to starboard and the trawler begins to swing in a great arc. The fishermen have to move quickly to surround the giant bait ball which is so near the surface...

The cables stretch and screech as the seine net closes. The winch gets hot as it reels in the steel line. The catch is secure now. The sardines have no downward escape route. The net will contain well over 30 tonnes of fish!

On the surface, it's mayhem. The water seethes with fish. The nets, kept afloat by numerous yellow buoys, are set upon by the seals as they dive over them in their tens, twenties, thirties, and soon in their hundreds or more. The whole scene is lit by floodlights. From an arm of the bridge, John oversees the operation at the stern, but I can feel that he is undecided about something. In the beam of the floodlights he has spotted the backs of the seals as they regularly surface to breathe. Each has a sardine poking out on either side of its mouth and looks for all the world as if it is laughing, as if it knew that it had sneaked in and stolen part of John's catch.

The real riches

In the early hours of the day, the thick tresses of the kelp forest shroud the first rays of the sun from the diver. A mere few shafts of light break through the disquieting darkness. For our team, as it ventures into this forest, these giant plants create a mysterious, almost troubling atmosphere. But as Roger Horrocks, a former spearfishing champion and now stills photographer, and I go deeper and our eyes get used to the darkness, the anxiety of the first few instants is replaced by a feeling of security. This is how I have always felt when diving in the kelp forest. The last few metres of the descent through the algae brings us to the seabed, where gradually everything becomes clearer. From the ocean floor things seem quite different: the beams of sunlight through the kelp are akin to those that filter through the highest stained glass windows of a cathedral before flooding the chancel with light.

Because this forest also provides protection from the swell, small fry and young prawns can mature here and they gather in myriads of tight, translucent shoals. My underwater lighting picks out their blue or pale pink

Sunlight breaks through the darkness of the kelp beds (right): these cathedral-like forests can grow in water as deep as 30m (100ft) and flourish in nutrient-rich waters such as those where upwellings bring nutrients up from the depths or the waste products of a seal colony make their own contribution.

colouring. At night, abalone scan the surroundings and clamber over rocks in search of food. Nearby some lobster spurn the cadaver of a fish and a seal comes to watch me with wide, startled eyes.

A little further out, where the ocean becomes deeper, the forest disappears, as its stems cannot grow at depths greater than 15m (50ft). Other forms of life take over and the particularly abundant static fauna means that you will never see a single square centimetre of rock that is not smothered with some form of life – thanks, yet again, to the meeting of the two oceans.

Early winter

As autumn ends, one depression after another reaches South Africa from the south-west, bringing long ocean swells that batter the Cape Peninsula. Enormous waves crash onto the dark, rocky coastline with a deafening din. The air is noticeably cooler and the sea takes on its wintery grey hue.

As the wind strengthens the sea turns white and, offshore, violent rainstorms reduce visibility. Cape gannets wheel with the wind as their wings skim the crest of the waves without ever touching them.

Over the last few weeks the water, too, has become colder. The Agulhas Current has weakened and is no longer bringing enough warmer water to maintain the mild temperatures in the area. The aristocrats of the fish world such as the tuna or the marlin have gone back to their temperate winter quarters in the north, as have the mako sharks that normally follow them. In False Bay, the great white sharks have returned to patrol Seal Island, looking for the fur seals that become their favourite food during the winter.

This gradual cooling of the atmosphere spreads right across the south of the country. The cold water then goes north-east and stretches out in a tentacle which hugs the coast and probes each and every little cove. It is this new, cooler environment that attracts the sardines, for this is where they find their ideal water temperatures of between 15° and 19°C (59–67°F).

Perched on Danger Point, armed with a pair of binoculars, Hugh Pearson, the producer of our film, is watching the show. A few gannets soar up to scan the surface of the sea, bunch together, and then plunge down

Malgas Island (above): one of only three Cape gannet colonies off the coast of South Africa. "Malgas" is the Afrikaans word for gannet and over 35,000 pairs currently breed here each year.

Gannet fledglings (above): they are pushed out of the nest at the age of about three months and form adolescent groups for a week or so before they are fully fledged. They seem to amuse themselves by playing with stones and feathers, which may help them learn skills that will be useful in adult life, such as manipulating objects with their beaks.

A large gannet chick (above): at eight weeks, its dark, waterproof feathers have developed and its down has almost gone. Its parents' work is nearly done.

Cape gannet
Morus capensis

Gannets nest in large colonies, with turret-shaped nests packed densely together on the ground. Males establish a territory around August or September and then work hard to impress a mate by calling, head-shaking, and bowing. Elaborate courtship and greeting displays follow, with new pairs stretching their heads skywards and gently tapping their bills together in a behaviour known as bill-fencing.

Partners co-operate to build a nest out of the abundance of guano found on the island. Females lay a single egg which both parents take turns to incubate by wrapping their feet around it. High concentrations of blood vessels in the feet allow heat to be transferred between the adult bird and the egg. Upon hatching six weeks later the chick will be black, naked, and blind and weigh only about 70g (2½oz). Within just eight weeks, however, it will outweigh its parents and remain heavier than them until it fledges at 14 weeks; by then its wingspan will measure as much as 1.8m (6ft).

KEY FACTS

Distribution endemic to Africa; breeding colonies are restricted to six offshore islands in the south of the continent.

Largest colony Bird Island in South Africa has an estimated 80,000 pairs present during the breeding season.

Hunting technique these powerful fliers can dive from as high as 30m (100ft) at speeds of 100kph (60mph), catching fish with spectacular precision.

into the water. They emerge a few seconds later, sometimes with a sardine poking out on either side of their beaks, then fly off into the wind before stationing themselves above the shoal which they have just signalled by their presence.

Hugh can feel it's nearly time for the sardines to leave and to start their great journey north.

Bird Island: the great exodus

Battered by the winds, Bird Island is an important point in our film because of its huge colony of Cape gannets. It's a low island that rises from the ocean in Algoa Bay, some 60km (40 miles) east of Port Elizabeth. In addition to the gannets, there are African penguins, cormorants, and gulls.

Hugh and Justin arrive by helicopter, which is the easiest way to get here; indeed, the fact that there is no port means that large ships can't dock and the distance between Bird Island and the coast does nothing to encourage small boats to make the trip. The lighthouse and the pair of little cottages occupied by the two Parks Board rangers who patrol the island all year round are the only buildings.

For a cameraman, Bird Island is a unique experience. At its centre, seen from above, the red of the lighthouse thrusts through an enormous expanse of white – the plumage of the gannets. The helicopter pad is at the north of the island, a spot carefully chosen to be as far from the birds as possible. Even so, as they descend, Justin can make out not only the gannets but also numerous pairs of African penguins; one pair has made its nest less than 1m (3ft) from where the nose of the helicopter comes to rest.

Gannet parents take it in turns to go on fishing trips (right): while one hunts the other guards the nest. When the fishing bird returns, the pair ritualistically greet each other before feeding the hungry chick.

Despite the powerful down-draught created by the rotors, the birds don't budge; neither do any of the others nearby. Obviously strong winds are nothing new for them!

The rainwater from the previous night, combined with the bird's guano, produces an overpowering smell of ammonia that assails the nostrils. The atmosphere soon becomes unbearable; it takes several minutes for the filming team to get used to this new environment. But it doesn't matter: it's a magnificent sight. Birds, heads to the wind, fly above the colony and fill the sky. They remain perfectly still, in a state of complete weightlessness. The urge to stretch out an arm and touch the tips of their wings, some of which are no more than 1m (3ft) or so away, is almost

An adult gannet returns from a fishing trip (above): Somehow it manages to locate its own nest and partner among the chaos of the colony.

overwhelming. As they turn their heads slightly, their blue eyes are watching Justin. The camera captures them, first in a sweeping shot and then a close-up of the head, the eyes, the beak; next a detailed shot of the feathers on the wings or tails and finally the feet. For Justin this is an unprecedented opportunity to take some very close-up shots. The birds' deafening cries surround him so that he can barely hear anyone speak. Under foot, the pairs are so closely packed that you can count five or six in each square metre.

A little further on is the birds' runway – a strip of bare earth from which they launch themselves for take-off, running into the wind and beating their wings. On land, the island disappears beneath a blanket of feathers into which your feet sink silently as if you were walking on a fleecy mattress. Around the island, some of the gannets take a dip in the sea, plunging down practically horizontally, whilst others hover over the water, following the shoals of halfbeaks as they swim near the surface. They make a few quick, repeated dives, as if, for them, this were the dress rehearsal for the forthcoming assault on the sardines.

As it is April, the gannets are still settled on the island, but already you can spot the first signs of their great exodus for the north-east. They are going to gorge and grow plump on all the good food that nature offers before they start breeding; the oily flesh of the sardines provides marine birds with incredible reserves of fat.

One of the noisiest jobs in the business (above): producer Hugh Pearson arouses a lot of curiosity as he records the sound of a vast gannet colony. His hat protects him from droppings as well as from the sun!

The dolphins of the deep

Our diving team, in the inflatable boat skippered by Sijmon de Waal, has chosen Algoa Bay to look for common dolphins. Vic Peddemors is with us. His research is helping us enormously, as is his knowledge of dolphins. It's time for them to set out on their great crusade and they begin to patrol the vast waters of the ocean in serried ranks, looking for sardines. Overhead, hundreds of gannets follow them. Who passes on the news to whom? Do the birds show the dolphin which way to go or do the dolphins draw the birds along in their wake? It's a question to which no-one really knows the answer yet.

But we do know it's the common dolphin that holds the key to the way the hunt operates. The dolphins locate the sardines at depth, using their sonar system, and, being excellent divers, can then separate a part of the school from the main body and bring it to the surface. The sardines regroup and form a bait ball which has no protection other than staying in the bunch, for the assailants have no idea which individual to target.

Common dolphins "porpoising" as they move at speed during a hunting trip (above): these dolphins are highly opportunistic and when the run is not available they will take more dispersed sardines, anchovies, squid, and other pelagic marine life.

Common dolphins (above): they can also hunt at great depths, taking squid near the sea bed.

Common dolphin
Delphinus delphis & Delphinus capensis

Common dolphins are widely distributed throughout all tropical, sub-tropical, and warm temperate seas, but there are actually two species: the short-beaked *Delphinus delphis* is found in offshore waters and occurs frequently in the eastern tropical Pacific, while the long-beaked *Delphinus capensis* occurs more in coastal waters, and is the species found off the South African Cape. There appears to be little interaction between the two groups.

The movements of common dolphin pods are known to correlate with seasonal shifts in prey abundance, and the population that feeds off Cape Agulhas is no exception – as many as 20,000 individuals are thought to arrive each year as the sardines begin their run. They locate their food using echolocation. A focused beam of high-frequency clicks is emitted from the dolphin's head. These clicks travel through the water until they encounter an object such as a school of fish and are bounced back. The dolphin's teeth are arranged in such as way that they work as an antenna, receiving the incoming sound and making it easier to pinpoint the exact location of the object.

KEY FACTS

Social grouping seen in pods of up to 2000 individuals, although groups of 10–500 are more usual.

Swimming speed normally 8–11kph (5–7mph), but have been recorded at 47kphr (29mph) in pursuit of food.

Distance can move up to 240–320km (150–200 miles) in just 48 hours.

The dolphins then repeatedly dive and swim under the bait ball, which they nudge up to the surface. Here the sardines become vulnerable to aerial attack from the gannets, while any individual that breaks free from the school falls victim to the first passing dolphin.

In the confusion, the largest of the predators now makes its entrance: a Bryde's whale appears from nowhere and charges into the bait ball, jaws wide open; it nets its catch in its gigantic gullet, then silently disappears into the abyss, leaving only a white trail of bubbles and scales behind.

Common dolphins feeding at the surface (next page): here the feeding activity is less intense than when they are corralling a bait ball.

The great hunting pack

As the migration continues, a third great predator which has been stealthily following the shoals of sardine appears: the copper shark.

If you ask a child to draw a shark, this is probably the one he will draw. With its pointed snout, mouth set neatly under the head and well-proportioned fins, there is nothing unwieldy or out of proportion about it, unlike the white shark or the tiger shark. Neither does it have the misshapen head of the hammerhead or the over-long tail of the fox shark. It is the epitome of elegance within the shark family.

Copper sharks live at great depths, in the cold waters of the south. And like the other predators, they join in the fun now it's here. During the sardine run, any underwater dive, even one some distance from the dolphins and the gannets, will show you that the sharks are there.

Copper shark
Carcharhinus brachyurus

Also known as bronze whalers or narrowtooth sharks, these stream-lined predators are found primarily in offshore waters along continental margins, although they often enter large coastal bays where they can be seen feeding on fish within the surf zone. This can obviously bring them into close contact with humans, and attacks have been recorded. Although there are no confirmed fatalities, this large and aggressive shark is considered to be dangerous to humans.

Sharks have limited success hunting shoaling fish on their own, as they lack the speed and co-operative strategies of dolphins. Left to themselves, sharks seem to be restricted to feeding on stragglers at the edges of the shoal. However, when a bait ball has been formed by a pod of dolphins, the predators appear to hunt in a mutually co-operative manner. The air-breathing dolphins work together to contain the swirling ball of fish, whilst the sharks slice through the shoal from below, pinning the sardines up against the surface of the water and cutting off their escape route into the depths. In this way a large ball of fish can be separated from the main shoal and all but eliminated in a matter of minutes.

KEY FACTS

Length over 3m (10ft)

Weight up to 300kg (660lb).

Distribution occurs in most warm temperate waters in the Indo-Pacific, Atlantic, and Mediterranean.

A copper shark in the shallows (right): it tunnels through a bank of sardines, edging them towards the surface.

Five, ten, perhaps fifteen even before you get down to 10m (33ft). Once you reach 15m (50ft), there they are, circling around you.

They hunt in a different way from the devious white shark, which attacks seals from below, or from the lightning-fast mako, which homes in on its prey at something like 60km (40 miles) an hour. The copper shark has signed a pact with its traditional enemy, the dolphin. It waits for the dolphin to serve the sardines on a platter, in the shape of a fine bait ball of fish trapped on the surface. A three-dimensional attack, which would have been impossible at depth, is now easy. Mouths wide open, the sharks swim up under the bait ball, then snap their jaws shut, trapping several fish between their teeth. This great marine hunting pack will keep together until journey's end, hunting down the sardines into the shallow waters of the north.

The calm

Our journey north continues. This morning the dew has strung out its silvery pearls all along the Transkei coastline. It's going to be a fine day,

the first for several days. We are near the small town of Wavecrest, at the mouth of the river. Diving conditions are poor. The dominant north-easterly of the last few days has caused an upwelling which has pushed up waters that are cold and murky, but rich in nutrients. Even the birds are having trouble finding fish. They're diving indiscriminately, often making the mistake of thinking that a line of tidal current is the dark streak of a shoal of sardines at depth. They surface looking flustered, with no sign of a fish in their beaks. They're using up a lot of energy for nothing.

It's the same for the dolphins. You can sense that they are out of action. Small, fragmented groups of a few hundred individuals have replaced the great squadron of 5000 beasts of the previous days. They're sluggish and from our little boat we can clearly see the calves taking advantage of a few moments' rest with their mothers. Assistant producer Joe Stevens takes the opportunity to make some sound recordings. He has set up an HF microphone on a bamboo pole which he dangles over the water as a shoal of dolphins passes. The quality of his recording is

Fur seals surfing are a common sight off the Southern African coast (above): close to the shore they can make a fast exit onto the land if their principal predator, the great white shark, is on the prowl.

excellent, as apart from the waves lapping the sides of the boat there's not a single sound. A few gannets dive past the floating microphone and the smile on Joe's face each time a bird hits the water tells me that he is pleased with the result.

About midday the sea is at its calmest and from time to time we come across a wide stretch of water where the surface has been smoothed out by the oil left in the wake of the sardine. The distinctive smell we catch as we pass tells us that the glittering little fish are there, deep down.

The neck of the funnel

It's now June. At this time of year the sun no longer lights up the waterfall that cascades down 40m (130ft) from the top of the cliffs. In this light you can see only the arch. The sea spray that floods through it adds to its ghost-like appearance. Waterfall Bluff is a compulsory stopping point on the sardines' migration route. The little bay at the foot of the cliffs traps the cold water of the coastal counter-current coming up from the south-west. The sardines that have got this far can rest for several days before continuing on their way north. What they need now is a new surge of cold conditions and the currents and winds which will carry them along the final stretch of their journey.

Off Mkambati, the Agulhas Current is less than 11km (7 miles) out to sea. Anything that rides this cold counter-current is necessarily very near the coast. On the surface, freighters on their way up to Durban sometimes hug the coast to within just 2–3km (1–2 miles), travelling more quickly than if they had chosen a route further out to sea. The humpback whales whose migration toward their birthing grounds coincides with the sardine run

Waterfall Bluff (above): the last resting place for the sardines before they commit themselves to the long trip up the Natal coast. At this point the continental shelf is so close to the shore that it forms a narrow funnel from which the fish have no escape.

have also realized that it is easier to swim close inshore, although they take a route much further out when they come back south with their calves.

Wind over the Transkei

Our base camp is set up in the Mkambati Nature Reserve, a magical spot of which we have high hopes, although the cool air forces us into warm jackets and we have had to bring everything from Natal – diving equipment, compressor, cameras, fuel reserves, spare parts for the boat engine, and enough food for several days; it is not easy to get to Mkambati by road, especially since the heavy rain of the last few days has made the route all but impassable.

We are the only guests here. The Transkei Region administers the reserve as best it can, given the lack of funds. This former leper institution and the land around it have been turned into a nature reserve with the intention of developing the local economy. Unfortunately, it is so run down that few tourists come. The Victorian lodge and the stone cottages that surround it are basic and poorly maintained. The plumbing leaks and the water out of the taps is dark brown. I've even taken a reading of more than 60 volts in the bathwater...

But for us, this is a strategic spot that allows us to watch the sardines swim extremely close to us. The small bay at the mouth of the Mkambati River can shelter inflatable dinghies when the weather is rough, though we will still have to fight our way out to the sea through the high rollers which batter the little estuary.

This morning we are not going to sea. For the past two days the west wind has been blowing hard. The sea is white and the swell is still rising.

A school of sardines (above): for once there seem to be no predators around and the sardines can proceed calmly on their journey.

The wind-swept rain has reduced visibility out to sea, but near the coast we can see a non-stop procession of Cape gulls as they skim the surface of the water in small groups. This west wind is a golden opportunity for us as it will bring the clear, offshore water in to the coast and should give us good conditions for filming underwater. We're just at the crucial point between the north-east winds which bring the right temperatures for the sardines and the west winds which give us good visibility.

We cross our fingers.

The next day the weather improves and our first ocean foray is a complete surprise. The water is unusually clear for this coast where a great number of rivers flow into the sea. On the sonar of our inflatable dinghy, Sijmon can see fish all around. True, it's not a dense shoal, but it extends over a huge area from the coast and is several kilometres long.

We make our first dive a few hundred metres out and go to a depth of 15–17m (50–55ft). Visibility is in excess of 15m (50ft) and the thermometer is showing 17°C (63°F). The sardines are indeed here, and what we see underwater is exactly what the sonar had shown on the screen. As regular as clockwork, a shark with its razor-sharp teeth thrusts its way through the shoal.

A volcano on the ocean's surface

The following morning the swell has dropped. The air is clear and visibility is good. We scan the horizon. A few dolphins appear to the front of the boat, then immediately dive again with their tails to one side as if they had suddenly decided to change direction at the last minute. A small group of Cape gannet cackles as they follow. The dolphins are on the lookout for sardines, there's no doubt about that. This morning we might see something spectacular. The conditions are perfect and, given the bad weather over the last few days, all the predators are hungry.

At about 11am, we are sailing between Waterfall Bluff and Port Saint Johns. Our radio crackles to life. It is Erick Weber, our microlight pilot alerting is to some promising action, several kilometers to the north. As we steam to the spot, the horizon becomes heavy with birds, and the dolphins, huge numbers of them, cut through the waves in long leaps as if they are off to an important meeting.

On the horizon an enormous white ball is forming – a massive concentration of gannets. The boat is still making brisk headway. Using two elastic straps I fasten the camera, a faithful old Arriflex Super 16 in its Subspace underwater housing, onto a box at the back of the boat. I use this instead of HD to take slow-motion pictures underwater and I can already imagine the film spiralling prematurely off its spool inside the case – the cameraman's perpetual nightmare. There are a few hundred metres more before we reach our destination. Sijmon, of course, hasn't slowed down and I'm having difficulty keeping on my feet so that I can see the action.

The predators move in

We are beginning to see the dolphins as they waltz and twirl on the surface. The birds are now diving in their hundreds, aiming at a specific point in the sea. The white water is seething with them. From our vantage point, all this activity looks like a volcanic eruption and the birds are the bombs of lava falling back down into the heart of the crater. At the edge of the stage, the gannets, in their hundreds, break through the surface of the water like a balloon shooting up from the depths. Each and every one has a sardine in its beak. On board the dinghy, the excitement is at its height. We are getting the equipment ready for the dive and I turn on the camera. The boat has stopped. Close up, the spectacle is even more dramatic. Hundreds of dolphins have trapped a huge bait ball on the surface. At its centre, copper sharks, swimming erect, snap up the sardines between their grinning jaws. In their frenzy, the sharks beat their tails on the surface of the water, scattering to the winds victim after victim which are immediately caught by the gannets. And there are thousands of them in the air, taking turns to swoop down on the sardines. The birds hit the water so near us that we start to worry about them puncturing the dinghy.

Gannets attack bait balls in relays (above): while some dive for the fish, others hang back to digest before taking another turn. Although this sounds orderly, the activity around a bait ball is frenzied – gannets create a huge racket as they strive to grab as many sardines as their bellies can carry.

A moment in the madness (right): a gannet takes advantage of the rich pickings that a dolphin has nudged its way.

Savagery beneath the surface

At last we are ready. Roger Horrocks and I put on our gear. Camera in hand, we make a backward flip and dive... into a war zone. Gannets slice through the surface, each creating a dull thud akin to the shell-fire from a World War 1 150-pounder. I'm amazed that we've lived to tell the tale! Visibility is not good, a few metres at most. All around us the water is foaming with billions of bubbles trailing after the birds as they plummet down. The raucous cries of the dolphins tell me that there are hundreds of them within striking distance too. As they rush up from the seabed, some of them brush past me, and I can't even see the sharks that must be in the thick of the fracas. Behind me, Roger is using his camera as best he can to fend off the sharks, who are getting a little too inquisitive.

You have to dive beneath the bait ball of fish to be able to see more clearly. As we go on down further, the water turns bluer and lighter, until visibility is up to a good 15m (50ft). This is perfect. Just above me, a new sight meets our eyes. Numerous dolphins are taking it in turns to keep the sardines near the surface, while hundreds of sharks force their way through the group from below, opening up tunnels of light as they move. From above, the bait ball is split from top to bottom by the long trails of bubbles left by the gannets.

The sardines do not know which way to turn. They swim a few metres in one direction, only to turn abruptly in another. Who has given the orders? Who is their leader? No-one knows. Some 15m (50ft) down, at the end of the trail of bubbles, the gannets glide on their way from the speed of the dive, and then beat their wings vigorously to propel themselves towards their prey. A quick forward thrust of the neck and the gannet has yet another sardine forming the smile across its beak.

Beneath the surface, it's pandemonium. The bait ball is clearly losing the battle: the dolphins and the sharks are attacking faster and faster, as if they want to have done. Despite the poor visibility, I venture into the thick of the action. There are so many fish, I can't see a thing in front of me. They swim so fast I quickly get giddy. I've lost all my bearings, both in time and in space. Copper sharks brush past me, each clutching five or six sardines in its teeth. A few of the little fish escape; their severed bodies quiver for a few seconds before being scooped up by another shark. The water is red and oily at the same time. My pressure valve is slimy and my camera almost slips out of my hands. The sharks' jaws swim past so close to me that I bury my hands inside the handles of the watertight casing. I don't want to risk being bitten. It does happen: I remember a dive in murky water with Tony White, the English photographer who was bitten by a copper shark sniffing around, looking for sardines...

Gannets diving for sardines (right): Gannets don't just pluck sardines from the surface of the ocean. They can "swim" underwater, using their wings to propel them, and have been seen at depths of 18m (60ft).

Shark netting (above): this is a controversial way of protecting beaches used for swimmers and surfers. Advocates of the system say that the carcasses provide scientists with measurements of the sharks, diet samples and other important data.

Less than an hour later, all that is left is a tiny bait ball which is of no interest to anyone. I can see the sardines' distress in their large, round eyes. Lost in this cruel sea, they now take advantage of the calm after the storm to scatter at speed and sink back to the bottom of the ocean. The sharks drift gently, weighed down by their bloated stomachs. The calm sound of the dolphins' breathing has replaced their shrill whistling and, at surface level, all I can see of 3000 birds are their webbed feet as a rain of silvery scales flutters serenely down into the deep.

The big catch

About 40m (130ft) down, the sardine survivors have regrouped in the thermocline – the boundary which marks the meeting of the warm surface waters and the cold bottom waters – and are continuing on their way north-east. The next day the first glimmers of dawn reveal the presence of some shadows on the beach at Port Edwards. For Bobby Naidoo and his team, it's nearly time for the great encounter.

Bobby belongs to the small band of professional fishermen who have a licence to cast their seine nets from the beaches. He's been on the lookout for several days now, just like the other fishermen and indeed the entire local population. The arrival of the sardines on these beaches is a major event, which the officials of the Natal Shark Board try to predict by making regular reconnaissance flights, following the fishes' progress along the coastline. Scientists must predict when the first shoals will arrive so that they can take in the 45km (30 miles) of anti-shark nets that protect the beaches. These nets were installed after five people died as a result of shark attacks between December 1957 and Easter 1958. Since then, they have proved their worth as a shark attack deterrent, with the number of incidents on the protected beaches declining dramatically. Yet leaving the nets in place when the sardine run arrives would have drastic consequences for the shark population, which would find itself caught in the trap, as happened in July 1995. I remember the panic on the beaches when the dead bodies of numerous sharks had to be removed, well out of sight of any onlookers, and the nets hauled in as quickly as possible to prevent another massacre.

It is 7am and the sun is rising over the coastline. The various teams of fishermen are scanning the sea beyond the huge waves. They want to be the first to spot the black, restless mass of sardines. The first boat to get onto the water will scoop the catch. Suddenly, Bobby's mobile rings. It's a call from one of his spotters, waiting upstream along the coast, warning him that a huge "pocket" of fish is just about to arrive.

Bobby's boat is ready to leave the beach, its net ingeniously stowed on board. The men will have to work quickly to surround the ball of fish as it is

Predators surround a bait ball (previous page): bait balls may be different shapes and sizes, but they are all attacked by predators in the same way.

pushed within reach by a particularly strong swell. It's fairly rare to get the chance to let the net out. You have to wait for all the right conditions to come together – wind, swell, and visibility – without forgetting that the seine net has to be deployed on a sandy bottom where there are no rocks to snag it as you haul it in again.

A few diving gannets attract Bobby's attention. Nearly every bird is surfacing with a sardine in its beak. The beach suddenly looks like the starting grid of a Formula 1 race. All the skippers of the various crews dash into their boats, ready to set off first. Bobby, too, immediately starts to manoeuvre. The coxswain jumps on board and fires the engine with the propeller still in the sand. Six men slide the little craft to the water's edge and then let go of it as it breasts the swell. On the beach, the long rope that will be used to haul in the net, uncoils at speed. Keep your hands and legs well clear! Bobby's team is in the lead as the little boats break through the big rollers which could capsize them. The skipper pays out the net, which now forms a perfect circle and imprisons a large section of the pocket of fish. A second boat arrives and it too uncoils a seine net around the shoal.

On the beach, Bobby's 30 men, who all sport a red T-shirt with the logo "First Light", haul on the hemp rope connected to the net, while a winch attached to the front of a Land Rover pulls in the other side; they have to trap the fish as quickly as possible before any sardines find a last way out.

A few metres away, another team of fishermen starts up a dance to their own chanting as they get ready to haul in a rope. This group is rigged out in blue T-shirts emblazoned with *Sardinops sagax*.

Sardine fever

The news that the sardines have arrived has already been broadcast by East Coast Radio, which has even created a hotline so that enthusiasts can follow the sardines' progress "live". Large numbers of cars are making their way to the beach and are soon blocking access to it. Only the fortunate owners of powerful 4x4s can avoid the long, difficult walk across the sand to reach the show being put on by the fishermen. Dozens and then hundreds of onlookers arrive, bucket in hand. And this is why. As soon as the sardine run is announced all the locals come along, to be a part of this traditional way of fishing in the hope of gleaning a few sardines from the nets as they come up.

The waves are now pushing Bobby's net towards the beach. The ropes strain in the violent backwash, then slacken as the swell pushes forward. The heads of the sardines poke from the black bulk of the swollen net as they try to wriggle through the mesh.

The net is on land now; to the seaward side of it, the fishermen are hammering wooden pickets into the sand to prevent it rolling backwards if it is caught by the backwash. This is a dangerous manoeuvre. All it needs is for a rope to snap and a fisherman could find himself engulfed in sardines... A thick, viscous oil slick oozes out of the net as it is hauled up the sandy beach.

A few sardines have escaped but they quickly fall prey to the onlookers, who don't hesitate to hurl themselves into the water to catch a glittering little fish. The survivors have lost most of their scales, their bloodied gills

Big-scale trawling

The pelagic purse-seine fishery is big business in southern Africa, yet in recent years the sardine industry has suffered huge losses. In Namibia the fishery opened during the early 1950s and immediately began netting huge quantities of fish, with catches rising from roughly 200,000 tonnes to a maximum reported catch of 1.4 million tonnes by the late 1960s. However, this level of harvesting was simply unsustainable and a sharp decline set in, until by the mid 1990s catches were varying from only 25,000 tonnes to almost nothing, and the fishery all but collapsed. Since then stricter controls and better management plans have been put in place, but the sardine stock in Namibia is still under threat.

Today South Africa's sardine catch is only 10 per cent of what it was 12 years ago, but it at least is showing signs of recovery due to good management, based mainly on a large reduction in fishing activity. Sardines harvested by purse-seine vessels still supply the greatest tonnage of fish landed each year in South Africa, but strict quotas determined annually by the Fisheries Authorities are in place to prevent another collapse.

Purse seiners (above): they fish around the sardines' breeding grounds on the Agulhas Banks throughout the year. Their high-tech equipment enables them to locate shoals efficiently.

are full of sand, and the twitching of their body heralds their imminent death – but sardine fever prevails and catching any fish is important, however woeful its condition.

"Shark! Shark!" Bobby yells. As the fishermen stand, braced, in the water, the dorsal fin of a copper shark, attracted by the oil slick, comes into sight through the gap between their legs. The shark throws itself at the seine net and stiffens as its jaws bite at the mesh, tearing open a yawning hole. It snatches some sardines before disappearing into the murky water. Every one has dashed out of the water, but within a few minutes feverish activity starts again.

The full net augurs a good catch. The net must be emptied quickly; and less than two hours later there are a hundred or so 20kg (44lb) crates piled up on the beach. While some of Bobby's team are getting the boat and the net ready for a second trip, others are taking away the boxes of fish.

The sun is high enough for us film an underwater sequence. I negotiate a place in Bobby's boat, promising that I will not take up much room and that I will dive as soon as the net is run out over the school of sardines. I'll have to swim quickly and film the scene from below. I know, too, that I will have to get back to the beach alone and unaided, and that I'll probably get a pummelling from the swell.

The images of the net settling over the school of fish remain vivid in my mind. I can still visualize the few fish that are swimming free but desperate to stay with the group. They whisk away to join their captive brothers in the net before it closes up definitively on the sand.

Epilogue

For me, the sardine run is one of the most fascinating events in the animal world to film. Of the eight years I have witnessed, each has been different. Sardines are unpredictable. They can show up anywhere along the coast without anyone knowing for certain where to find them. The great difficulty for a filmmaker is being there at just the right time, as the bait ball comes to the surface only very briefly, rarely for more than an hour. You never get the right conditions for perfect filming; when the animals are there, the water isn't clear enough and vice versa. And how many times have I heard: "Oh, you should have been here yesterday... "

While the gannets on Bird Island and the sardine-hunting seals of the Cape Province obliged us, the sardines themselves and their predatory followers kept us in a state of nail-biting tension as the days and the

Less sophisticated fishing off the Natal coast (left): these sardines have been caught in a net dropped behind a boat in the shallows. For a couple of weeks almost every year the area is in the grip of sardine fever.

budget allocated to our filming ticked remorselessly away. We even had a helicopter on stand-by for part of the time, ready to sweep us away to wherever the sardines chose to arrive the moment we had word of them.

The sardine run can drive a filmmaker mad.

Not only that, but things seem to have changed over the last few years. Tiger and black-tip sharks, which normally desert the Natal coast in winter when the temperature drops below 21–22°C (70–72°F), are now there all year round. In 2006 and 2007, the temperature did not drop below 23°C (73°F), whereas in previous years it had been down to 17–18°C (63–64°F). In 2007, for the first time, two humpback whales, en route for their winter quarters in the warm waters off Mozambique or Madagascar, stopped off on the Aliwal Shoal on the Natal coast to sleep – another sign that the water was warmer than it used to be.

In the absence of this tongue of nutrient rich cool water, will the sardines continue to be lured this far north? My description of

going fishing with Bobby is taken from one of my experiences in 1999. The sardine run has been a major event in the fishermen's life for many years, but in recent years, very disappointing catches have been recorded. The year 2007 for example was one of the poorest for decades! Bobby no longer gets the phenomenal catches he used to.

A further concern is the fact that many of the sardine fishing vessels that used to be based on the Atlantic coast have now settled in Mossel Bay, east of Cape Town. Some 80 fishing boats are hauling out a total of 800 tonnes of fish a night with the same quantity of bycatch – unwanted small sardines and other marine life that is caught up in the commercial nets, then dumped back in the sea, dead or dying, with devastating effects on the entire food chain.

It is, of course, too soon to assert loud and clear that the sardine run phenomenon is in decline, but – given its unpredictability over the last few years – it is nevertheless possible that we are witnessing a major alteration in the natural world.

Sunset over Malgas Island (above): these Cape gannets are only one of several predatory species that face an uncertain future if the phenomenon that is the sardine run ceases to exist.

The authors

Brian Leith – Executive Producer, *Nature's Great Events*

Brian Leith has been a producer/director for over 20 years, working in the BBC Natural History Unit as well as running an independent production company. His films have won numerous gongs – including two Best of Festival awards, a Royal Television Society award, and a Prix Italia. He has written a regular column (GreenPiece for *The Listener*) and published a book (*The Descent of Darwin*, for Collins).

Mike Holding – Principal Cameraman, The Great Flood

Mike Holding is an accomplished director and cinematographer of international television wildlife documentary films and docu-dramas. AfriScreen Films was formed by Mike together with producer Tania Jenkins over a decade ago. Together they have successfully worked on many blue-chip documentaries, and have produced two popular natural history films, *A Wild Dog's Story* and *Swamp Cats*, which have won several prestigious international awards.

Joe Stevens – Assistant Producer, The Great Feast

Joe joined the BBC NHU five years ago, previous to which he was a winner of the prestigious Rolex International Dive Scholarship. Since joining the NHU, he has been involved in a huge range of projects including *Elephants of Samburu*, *Dive Belize*, and the BBC's acclaimed *Galapagos* series. Joe has also produced and directed a number of short films and was involved with writing the book that accompanied the *Galapagos* series.

Amanda Barrett – Researcher, The Great MIgration

Amanda Barrett has worked with film maker Owen Newman for 20 years. Their films for the BBC's Natural History Unit have won many international awards and been praised by the wildlife film making industry for their revealing and intimate portraits of animals. In recent years, they have specialized in working with cats in the wild that have been previously impossible to film. Few film makers alive today can rival their experience in East Africa.

Karen Bass – Series Producer, *Nature's Great Events*

Karen Bass is a multi award-winning producer. Among her award-winning films are *Pygmy Chimp – The Last Great Ape*, the first film to be made about bonobos of the Congo, and *Crocodile Wildlife Special. Andes to Amazon*, a landmark series about the natural history and extreme landscapes of South America, *Jungle*, investigating the world's rainforests, and *Wild Caribbean* are among Karen's most recent successes.

Justin Anderson – Producer, The Great Melt

Justin has been working for the BBC Natural History Unit for seven years. He first travelled to the High Arctic in 2003 to present the BBC Radio 4 programme *Among Arctic Wolves*. He then went on to work on the award-winning *Planet Earth* series, filming in Borneo, Madagascar, America, and Australia for the episodes on "Caves", "Deserts", and "Ice Worlds". When not exploring the icy north he likes to be writing at home, in the warmer climes of Bristol.

Jeff Turner – Producer, The Great Salmon Run

Jeff has been making documentary films with his wife Susan for the past 22 years. They have directed, shot, written, and produced more than 25 documentaries for the BBC, CBC, PBS, Discovery, and Animal Planet channels and have also won numerous awards for their work. Jeff and Sue have a strong connection to bears, having produced six different films on bears since 1991. For the past 17 years Jeff and Sue have been working exclusively for the BBC.

Didier Noirot – Cameraman, The Great Tide

Didier has been a diver, underwater photographer, and cameraman for over 30 years. For 12 years he was the underwater cameraman for the great Jacques Cousteau on board the *Calypso*, which took him all over the world. He has filmed and headed up underwater cinematography on over 20 documentaries, including several on the sardine run. He was also the underwater camerman on *Sensitive Sharks*, *Blue Planet*, and *Planet Earth* for the BBC.

Research by Louise Emerson and Ed Charles

Find out more

The following is a brief list of suggestions for those seeking more information on the great events covered in this book.

The Great Flood

Bailey, A., *Okavango: Africa's Wetland Wilderness*, Struik Publishing, 2005

Harvey, C., *Beyond the Endless Mopane*, Southern Book Publishers, 1997

Johnson, P., & Bannister, A., *Okavango: Sea of Land, Land of Water*, Country Life Books, 1984

Lanting, F., *Okavango: Africa's Last Eden*, Chronicle Books, 1993

Mendelsohn, J., & el Obeid, S., *Okavango River: Flow of a Lifeline*, Struik Publishing, 2004

Ross, K., *Okavango: Jewel of the Kalahari*, BBC/Southern Book Publishers, 1987

The following DVDs also cover the Okavango and its flood:

Okavango – Jewel of the Kalahari, Partridge Films/BBC

Planet Earth, Episode 1, BBC NHU

Okavango: Savage Paradise, National Geographic

Swamp Cats, Afriscreen Films/BBC NHU

Haunt of the Fishing Owl, Tim Liversedge/Partridge Films

The Great Feast

Ford, C., *Where the Sea Breaks its Back: the epic story of early naturalist Georg Steller and the Russian exploration of Alaska*, Little, Brown, 1966

Miller, M. *Alaska's Southeast: Touring the Inside Passage*, Globe Pequot Press, 11th edition 2008

O'Clair, R. M., Armstrong, R. H., & Carstensen, R., *The Nature of Southeast Alaska: a guide to plants, animals, and habitats*, Alaska Northwest Books, 1992

Also try these websites:

www.adfg.state.ak.us

www.alaskafisheries.noaa.gov

www.afsc.noaa.gov

www.marinemammal.org

www.whaletrust.org

The Great Migration

There are hundreds of publications on the Serengeti ecosystem, ranging from glossy picture books to indispensable but more esoteric research papers. Some of the most accessible and informative are:

Estes, R.D., *The Behaviour Guide to East African Mammals*, University of California Press, 1991

Reader J., *Africa: A Biography of a Continent*, Penguin, 1998

Reader J. & Croze, H., *Pyramids of Life*, William Collins & Sons, 1977

A taste of what it's like to work in East Africa is given by Robert Sapolsky's brilliantly entertaining *A Primate's Memoirs: Love, Death and Baboons in East Africa* (Jonathan Cape, 2001), while no reader can afford to miss the invaluable documentary history of our relationship with the world around us in Jared Diamond's international bestseller *Collapse: How Societies Choose to Fail or Succeed* (Allen Lane, 2005).

The Great Melt

Gulliken, B., & Svensen, E., *Svalbard and Life in Polar Oceans*, Kom Forlag, 2004

Kovacs, K. M. & Lydersen, C. (eds.), *Birds and Mammals of Svalbard*, Norwegian Polar Institute, 2006

Paine, S., *World of the Arctic Whales: Belugas, Bowheads and Narwhals*, Sierra Club Books, 1997

Rosing, N., *The World of the Polar Bear,* A& C Black, 2006

Sale, R., *The Complete Guide to Arctic Wildlife*, Christopher Helm, 2006

Stirling, I., photographs by D. Gurvich, *Polar Bears*, University of Michigan Press, 1988

The Great Salmon Run

Croot, C., & Margolis, L. (eds.) *Pacific Salmon Life Histories*, University of British Columbia Press, 2003

Forbes, N., Jay, T., & Matsen, B. *Reaching Home: Pacific Salmon, Pacific People*, Alaska Northwest Books, 1995

McAllister, I., *The Last Wild Wolves: Ghosts of the Rainforest*, University of California Press, 2007

McAllister, I., & McAllister, K., with C. Young, *The Great Bear Rainforest: Canada's Forgotten Coast*, Harbour Publishing, 1998

Quinn, T. P., *The Behaviour and Ecology of Pacific Salmon and Trout*, American Fisheries Society, Maryland, in association with University of Washington Press, 2005

The Great Tide

Aitken, A., *Sardine Run: The Greatest Shoal on Earth: a field guide to the sardine run, sharks and other marine life on the eastern coast of South Africa*, Natal Sharks Board, n.d.

Jacana Maps, *Garden Route Guide*, Jacana Education, 3rd edition 2006

Peschak, T. P. & Velasquez Rojas, C. *Currents of Contrast: Life in Southern Africa's Two Oceans*, Struik Publishers, 2006

Also try these websites:

www.sardine-run.com

www.acsonline.org

Index

Acknowledgements

Nature's Great Events has been a big team effort. The experience of making the programmes forms the basis of the chapters of this book and behind every filming trip there was a large support team, without whom none of this would have been possible.

Production coordinators and production team assistants Loulla Charalambous, Clare Crossley, Katie Cuss, Katherine Gilbert, Gemma Greene, and Karen Le Huray were responsible for keeping us all on track. Production manager Anna Gol ensured, amongst many other things, that we stayed on budget. Peter Bassett, producer of The Great Flood and The Great Migration, worked closely with Mike Holding, Amanda Barrett and Owen Newman in shaping their filming projects, while Hugh Pearson, producer of The Great Feast and The Great Tide, worked closely with assistant producer Joe Stevens and underwater cameraman Didier Noirot. In addition to their day jobs the series researchers, Ed Charles and Louise Emerson, contributed to many of the chapter boxes and much else besides. Laura Barwick searched tirelessly for fresh images to illustrate our stories. Ailish Heneberry helped to get the book off the ground.

In addition there are numerous cameramen, location managers, scientists, and local people across the globe who provided technical and moral support. Their professionalism, knowledge, and photographic talents have enhanced this book as well as the series. To you all, our heart-felt thanks.

Karen Bass Series Producer

In addition, the following people and organizations helped to make the individual films and chapters possible:

The Great Flood

Tania "TJ" Jenkins, Richard Uren, Adrian Dandridge, Peter Bassett, Dave and Helene Hamman, Kai Collins, Clinton Edwards, Kelly Landen, Joel Mawire, the Government of Botswana, the Botswana Department of Wildlife and National Parks, Chitabe Camp, and Wilderness Safaris.

The Great Feast

Photography: Paul Atkins, Barrie Britton, Tom Fitz, Shane Moore, David Reichert, Warwick Sloss, Gavin Thurston, and Simon Werry; field assistants and location crew: Pamela Bendall, Brian and Michelle Gundaker, Kent Hall, John Hyde, Michael Mauntler, Scott and Dustin McLoed, Randy Miller, Stephie Olsen, Angela Smith, Jason Sturgis, Dave Svobodny, Igor Valente, Bill Weeks, and Don Wilson; scientists and consultants: Ralph Dunlop, Mark and Debbie Ferrari, Mark Hipfner and the Simon Fraser University research team, Jamie Ross and family, and Jan Straley. Thanks also to Temsco Helicopters, Ministry of Environment, B.C. Parks, West Coast Helicopters and Joe Stevens' personal thanks to Leticia Valverdes.

The Great Migration

Behind every film crew there are many people who are instrumental in the success of the finished product. Our heartfelt thanks to Janice and Richard Beatty, Gerald Bigurube, David Breed, Sarah Durant, Aadje Geertsema, Marleen Koppen, Margaret Kullander, Colin McConnell, Hamisi Masawe, Bernard Murunya, Pat Patten, Nigel Perks, Angus Simpson, Ben Simpson, Mike Watson, Paul White, Benoît Wilhelmi, and all the staff of Ndutu Safari Lodge. Without their expertise, willing help, warm friendship and sense of fun, our lives in the field would have been almost impossible. Thanks also to the rangers, wardens and researchers of both Ngorogoro Conservation Area and Serengeti National Park who gave us invaluable support, and the Tanzanian authorities for allowing us to work in their country. Not to mention an incalculable debt of gratitude to the birds and beasts in the Serengeti for their unwitting co-operation.

The Great Melt

Photography: Martyn Colbeck, Warwick Sloss, Simon Werry, Tom Fitz, Göran Ehlmé, and David Mckay; location manager Jason C. Roberts; and field assistants Steinar Aksnes, Arne Sivertson, Lasse Ostervold, Joe Barron, Heli Routti, Paul Beilstein, and Richard, Josee, Nansen, and Tessum Weber. Thanks also to MS *Polarhav*, Kjell Arild Hansen, Grete Eriksen, Line Forisdal, Lorents Lorent, Svenn Are Johansen, and Donni and Marilyn Wakowski

The Great Salmon Run

Ian and Karen McAllister and their children, based near Bella Bella. As founding directors of the Raincoast Conservation Society, Ian and Karen have been heavily involved with environmental and community work in the Great Bear Rainforest for over 20 years. Thanks also to scientists and researchers Dr Morgan D. Hocking in British Columbia, Anders Lamberg in Norway, and Professor Thomas P. Quinn of the School of Aquatic and Fishery Sciences, University of Washington; and to Harvey Humchit, First Nations Chieftain living in Bella Bella. Jeff Turner's special thanks to Susan Turner, his partner and co-producer, for her continuing advice, support and guidance.

The Great Tide

Photography: Justin Maguire and Simon Werry; field assistance and location crew: Roger Horrocks, Blue Wilderness Dive Expeditions, Mark and Gail Addison, and Eric Webber; scientists and consultants: Gwen Penry, Vic Cockcroft, and Debbie Young.

Photography credits

Key: a above, b below, c centre, l left, r right

Map/graphic designer: Martin Darlison at Encompass Graphics
Endpapers: Ian McAllister/Pacific Wild

Introduction

5al Adrian Bailey/Photolibrary Group, 5ac Francois Gohier/ ardea.com, 5ar Owen Newman, 5bl Paul Nicklen/ National Geographic Image Collection, 5bc Paul Zakora, 5br rogerhorrocks.com 2008, 6–7 NASA/Science Faction/Getty

Flood

Box background Adrian Dandridge, 12–13 Adrian Bailey/ Photolibrary Group, 14–15 Richard du Toit/Minden Pictures/ FLPA, 16 Paul A Souders/Corbis, 17 Martin Harvey/Corbis, 18–19, 20, 21 Frans Lanting/Corbis, 20l Patridge Films Ltd/Photolibrary Group, 22a Daryl Balfour/NHPA, 22b Bobby Haas/ National Geographic/Getty, 23a NASA/Corbis, 23b Adrian Dandridge, 25 Peter Bassett, 24–5, 26, 27 Kelly Landen, 28–29 Steve Bloom Images/Alamy, 30a Beverly Joubert/National Geographic/ Getty, 30b Dave Hamman, 31 Richard Packwood/Photolibrary Group, 32a, 32b Peter Bassett, 33 Dave Hamman, 34 imagebroker/ Alamy, 35 Frans Lanting/Corbis, 36 Anthony Bannister/NHPA/ Photoshot, 37 Paul Franklin/ Photolibrary Group, 38 Adrian Dandridge, 38–39 Frans Lanting/ FLPA, 40–41 Adrian Dandridge, 42 Theo Allofs/Corbis, 43l Frans Lanting/FLPA, 43r Adrian Dandridge, 44–5 Frans Lanting/ FLPA, 46–7 Richard du Toit/Minden Pictures/FLPA, 48 Martin Harvey/ Corbis, 49, 50–51 Dave Hamman, 52a, 52b Kelly Landen, 53 Marcus Wilson-Smith/Alamy, 54–5 Frans Lanting/FLPA, 56–7 Beverly Joubert/National Geographic/ Getty, 58 Clinton Edwards, 58–9 Peter Bassett, 60–1 Michael Poliza/Getty, 62 Tony Heald, 63 Mike Holding, 64–5 Michael Melford/Getty

Feast

Box background M I Walker/ Science Photo Library, 66–7 Francois Gohier/ardea.com, 68–9 Danny Lehman/Corbis, 70–1 Paul Nicklen/National Geographic/Getty, 72 both www.visibleearth.nasa.gov, 73 George Lower/Science Photo Library, 74–5 M I Walker/Science Photo Library, 76–7 Joe Stevens, 77 Bryan Gundaker, 78–9, 80–81 Ed Charles, 82 Roy Toft/National Geographic/Getty, 83 Barrie Britton, 84–5 Alaskastock/ Photolibrary Group, 85, 86a Barrie Britton, 86b Alaskastock/ Photolibrary Group, 87 Doug Allan/Photolibrary Group, 88–9 Michael Gore/FLPA, 90–91 Chris Newbert/Minden Pictures/Getty, 92–3 Flip Nicklin/Minden Pictures/FLPA, 93 Joe Stevens, 94l Fred Bavendam/Minden Pictures/FLPA, 94r, 95 Brandon Cole/ naturepl.com, 96–7 Paul Nicklen/ National Geographic/ Getty, 98 Cornforth Images/Alamy, 99 all Hugh Pearson, 100–1 Randy Miller and 102b Hugh Pearson, both Center for Whale Studies, 2008, taken under NMFS Permit #393-1772, 102–3 Francois Gohier/ ardea.com, 104–5 Mitsuaki Iwago/Minden Pictures/FLPA, 106–7 Richard Olsenius/National Geographic/Getty

Migration

All images © Owen Newman except: 116–7 Winfried Wisniewski/zefa/Corbis, 125b Kevin Schafer/ Corbis. 132 Paul A Souders/Corbis, 136b Owen Newman Ltd

Melt

All images © Jason C Roberts except: 158–9 Paul Nicklen/National Geographic Image Collection, 161–2 Doug Allan/naturepl.com, 164–5 Jim Brandenburg/Minden Pictures/FLPA, 166 Fred Hirschmann/Science Faction/Getty, 166–7 Norbert Rosing/National Geographic/ Getty, 169 Justin Anderson, 170–1 Darrell Gulin/Getty, 172–3 Doc White/naturepl.com, 172b Hanne & Jens Eriksen/ naturepl.com, 174 Martha Holmes/naturepl.com. 175a Paul Nicklen/National Geographic/ Getty 175b Sue Flood/ naturepl.com, 176–7 Norbert Rosing/National Geographic/ Getty, 178a Daniel J Cox/Getty, 178c Silvestris Fotoservice/FLPA, 178b Roger Tidman/FLPA, 179 Richard Herrmann/Photolibrary Group, 180–1, 182–3, 183b, 183c Justin Anderson, 184 Tom Brakefield/zefa/Corbis, 186a, 186br Flip Nicklin/Minden Pictures/FLPA, 186bl blickwinkel/Alamy, 186–7 Justin Anderson, both 189 and all 190 BBC, 191 Warwick Sloss, 196 S Jonasson/FLPA, 197 Steve Packham/naturepl.com, 198–9 George McCarthy/naturepl.com, 200l Owen Newman/Photolibrary Group, 200r Daniel Cox/ Photolibrary Group, 201b Justin Anderson, 202–3 Patricio Robles Gil/naturepl.com, 209c Theo Allofs/ Corbis, 209b Justin Anderson, 210–11 Andre van Huizen/Alamy

Salmon Run

All images © Ian McAllister/Pacific Wild except: box background Simon Werry, 212–3 Paul Zakora, 216 Neil Rabinowitz/Corbis, 217 Dan Lamont/Corbis, 218, 219 Paul Zakora, 220 Chris Cheadle/Getty, 221 Michio Hoshino/Minden Pictures/FLPA, 222–3 Peter Scoones/naturepl.com, 222l Michel Roggo/naturepl.com, 222r Brandon Cole/naturepl.com, 223l Darryl Leniuk/Getty, 223r Thomas Kitchin and Victoria Hurst/Getty, 223b Natalie Fobes/Corbis, 224–5 W. Perry Conway/Corbis, 227 Paul Zakora, 228–9 Brandon D Cole/ Corbis, 230–1 Flip Nicklin/Minden Pictures/FLPA, 232 Gerry Ellis/Minden Pictures/FLPA, 233 Michel Roggo/naturepl.com, 234–5 Anton Art/Alamy, 237 John E Marriott/Alamy, 242 Paul Zakora, 244–5 Gerry Ellis/Minden Pictures/ FLPA, 250–1 Chris Cheadle/Getty, 252a Eric Baccega/naturepl.com, 252b Hans Christoph Kappel/ naturepl.com, 253 SeaPics.com, 254–5 Brandon Cole/naturepl.com, 256a Gerry Ellis/Minden Pictures/Getty, 257 Paul Zakora, 262 both Natalie Fobes/Corbis 264b Paul Zakora

Tide

All images © rogerhorrocks.com 2008 except: box background SeaPics.com, 268–9 Image Source/Corbis, 270–1 SeaPics.com, 272–3 Justin Maguire, 274 Martin Harvey/Corbis, 275 Doug Perrine/naturepl.com, 276–7 SeaPics.com, 288r, 289 Justin Maguire, 290–1 Hugh Pearson, 292–3 Mark Carwardine/ naturepl.com, 294–5, 296–7 SeaPics.com, 298–9 Chris Fallows/ splashdowndirect.com, 300–1 both SeaPics.com, 302 BBC, 303, 304–5 SeaPics.com, 306–7 Doug Perrine/naturepl.com, 308 SeaPics.com, 309 Richard du Toit/naturepl.com

Author pictures

314al Richard Uren, 314ar Justin Anderson, 314cl Leticia Valverdes, 314cr Susan Turner, 314bl Owen Newman, 314br Rob Malpage